Student Solutions Manual for

The Statistical Sleuth

A Course in Methods of Data Analysis

Second Edition

Fred L. Ramsey
Oregon State University

Daniel W. Schafer
Oregon State University

DUXBURY
THOMSON LEARNING

Australia • Canada • Mexico • Singapore • Spain • United Kingdom • United States

DUXBURY
THOMSON LEARNING™

Assistant Editor: Ann Day
Marketing Manager: Tom Ziolkowski
Marketing Assistant: Mona Weltmer
Editorial Assistant: Jennifer Jenkins
Production Editor: Stephanie Andersen
Permissions Editor: Sue Ewing
Cover Design: Denise Davidson
Cover Illustration: Cici Man/Stock Illustration Source
Print Buyer: Micky Lawler
Printing and Binding: Webcom Ltd.

For more information about this or any other Duxbury products, contact:
DUXBURY
511 Forest Lodge Road
Pacific Grove, CA 93950 USA
www.duxbury.com
1-800-423-0563 (Thomson Learning Academic Resource Center)

Printed in Canada

10 9 8 7 6 5 4 3 2 1

ISBN: 0-534-38950-3

Contents

Chapter 1:	Drawing Statistical Conclusions	1
Chapter 2:	Inference Using t-Distributions	2
Chapter 3:	A Closer Look at Assumptions	2
Chapter 4:	Alternatives to the t-Tools	4
Chapter 5:	Comparisons Among Several Samples	5
Chapter 6:	Linear Combinations and Multiple Comparisons of Means	7
Chapter 7:	Simple Linear Regression: A Model for the Mean	8
Chapter 8:	A Closer Look at Assumptions for Simple Linear Regression	11
Chapter 9:	Multiple Regression	15
Chapter 10:	Inferential Tools for Multiple Regression	17
Chapter 11:	Model Checking and Refinement	20
Chapter 12:	Model Selection with Large Numbers of Explanatory Variables	25
Chapter 13:	The Analysis of Variance for Two-way Classification	27
Chapter 14:	Multifactor Studies	29
Chapter 15:	Adjustment for Serial Correlation	31
Chapter 16:	Repeated Measures	35
Chapter 17:	Exploratory Tools for Summarizing Multivariate Responses	37
Chapter 18:	Comparisons of Proportions or Odds	38
Chapter 19:	More Tools for Tables of Counts	38
Chapter 20:	Logistic Regression for Binary Response Variables	38
Chapter 21:	Logistic Regression for Binomial Counts	39
Chapter 22:	Log-Linear Regression for Poisson Counts	41
Chapter 23:	Elements of Research Design	42
Chapter 24:	Factorial Treatment Arrangements and Blocking Designs	42

Preface: To The Student

This solutions manual contains sketches of solutions to selected "computational exercises" that appear at the ends of the chapters in *The Statistical Sleuth* by Ramsey and Schafer.

There are no solutions to "Data Problems" here, but we offer a few words of advice about these. First, we recommend that upon completion of a data analysis problem you attempt to write a summary to communicate your results. See the "Summary of Statistical Findings" sections in the book for examples of statements used in conclusions. A well-chosen graphical display can also be very helpful in communicating the results. Realize that there are often dead ends in a statistical analysis. It is not helpful to provide your audience with a history of what you did. Instead, communicate what it is that was learned.

Check the *Sleuth* web site: www.statisticalsleuth.com for updates and corrections.

Good luck. Happy sleuthing.

Chapter 1: Drawing Statistical Conclusions

1.17 The difference between averages (A - B) in the observed outcome is 78.00 - 62.67 = +15.33 points. In the list that follows, there are three outcomes (nos. 1, 34, and 35) that have a difference as large or larger in magnitude as the observed difference. The two-sided *p*-value is therefore 3/35 = 0.0857.

Outcome No.	Guide A	A Average	Guide B	B Average	(A - B) Difference
1	53, 64, 68, 71	64.00	77, 82, 85	81.33	-17.33
2	53, 64, 68, 77	65.50	71, 82, 85	79.33	-13.83
3	53, 64, 68, 82	66.75	71, 77, 85	77.67	-10.92
4	53, 64, 68, 85	67.50	71, 77, 82	76.67	-9.17
5	53, 64, 71, 77	66.25	68, 82, 85	78.33	-12.08
6	53, 64, 71, 82	67.50	68, 77, 85	76.67	-9.17
7	53, 64, 71, 85	68.25	68, 77, 82	75.67	-7.42
8	53, 64, 77, 82	69.00	68, 71, 85	74.67	-5.67
9	53, 64, 77, 85	69.75	68, 71, 82	73.67	-3.92
10	53, 64, 82, 85	71.00	68, 71, 77	72.00	-1.00
11	53, 68, 71, 77	67.25	64, 82, 85	77.00	-9.75
12	53, 68, 71, 82	68.50	64, 77, 85	75.33	-6.83
13	53, 68, 71, 85	69.25	64, 77, 82	74.33	-5.08
14	53, 68, 77, 82	70.00	64, 71, 85	73.33	-3.33
15	53, 68, 77, 85	70.75	64, 71, 82	72.33	-1.58
16	53, 68, 82, 85	72.00	64, 71, 77	70.67	+1.33
17	53, 71, 77, 82	70.75	64, 68, 85	72.33	-1.58
18	53, 71, 77, 85	71.50	64, 68, 82	71.33	+0.17
19	53, 71, 82, 85	72.75	64, 68, 77	69.67	+3.08
20	53, 77, 82, 85	74.25	64, 68, 71	67.67	+6.58
21	64, 68, 71, 77	70.00	53, 82, 85	73.33	-3.33
22	64, 68, 71, 82	71.25	53, 77, 85	71.67	-0.42
23	64, 68, 71, 85	72.00	53, 77, 82	70.67	+1.33
24	64, 68, 77, 82	72.75	53, 71, 85	69.67	+3.08
25	64, 68, 77, 85	73.50	53, 71, 82	68.67	+4.83
26	64, 68, 82, 85	74.75	53, 71, 77	67.00	+7.75
27	64, 71, 77, 82	73.50	53, 68, 85	68.67	+4.83
28	64, 71, 77, 85	74.25	53, 68, 82	67.67	+6.58
29	64, 71, 82, 85	75.50	53, 68, 77	66.00	+9.50
30	64, 77, 82, 85	77.00	53, 68, 71	64.00	+13.00
31	68, 71, 77, 82	74.50	53, 64, 85	67.33	+7.17
32	68, 71, 77, 85	75.25	53, 64, 82	66.33	+8.92
33	68, 71, 82, 85	76.50	53, 64, 77	64.67	+11.83
34	68, 77, 82, 85	78.00	53, 64, 71	62.67	+15.33
35	71, 77, 82, 85	78.75	53, 64, 68	61.67	+17.08

1.19 Coin flips will not divide the subjects in such a way that there is an exact age balance. However, it is impossible to tell prior to the flips which group will have a higher average age.

1.21 There is no computation involved. This is, however, a sobering exercise.

1.23 The box plot should look a bit like the stem and leaf diagram in exercise #22.

1.25
- **a** No radiation median is 0; radiation median is 1.
- **b** Both distributions are positively skewed. The radiation group has a larger spread.
- **c** Over half the numbers in the set are = 0.
- **d** It is observational data, so a strict interpretation would say that causation cannot be inferred. But what else could it be?

Chapter 2: Inference Using *t*-Distributions

2.13
- **a** (Fish, Regular): Averages are (6.571, -1.143); SDs are (5.855, 3.185)
- **b** Pooled SD = 4.713
- **c** SE for difference = 2.519
- **d** d.f. = 12; $t_{12}(.975) = 2.179$
- **e** 95% CI from 2.225 to 13.203 mm
- **f** *t*-stat = 3.062
- **g** One-sided *p*-value = .005. Using the table in Appendix 2, locate the d.f.=12 line, and move across the line until the position of the *t*-statistic, 3.062. It is slightly larger than 3.055 so the table tells you that the one-sided *p*-value is slightly smaller than .005.

2.15 *t*-statistic = 9.32, with 174 d.f. Very convincing, indeed.

2.19
- **a** Average = -1.14; SD = 3.18; d.f. = 6.
- **b** SE = 1.20
- **c** 95% CI: from -4.09 to 1.80
- **d** *t*-statistic = -0.95; two-sided *p*-value = .38. [Using the table: $0.906 < 0.95 < 1.134$, so 0.95 is between the 80th and the 85th percentiles. The one-sided *p*-value is therefore between 1-0.80 = 0.20 and 1-0.85 = 0.15, and the two-sided *p*-value is between 0.40 and 0.30 (by doubling).]

Chapter 3: A Closer Look at Assumptions

3.21
- **a** One-sample t-test on differences (observed - expected) for the subset of umpires whose lifetimes were not censored (Censored = 0): t-stat = -0.987, df = 194, p-value = 0.32 (1-sided p-value = .16). A 95 percent confidence interval for mean life length minus expected life length: -1.6 years to 0.54 years
- **b** This might be a problem if the ones for whom data were unavailable tended to have died young. In any case, the available sample is not a random sample from the population of all umpires.

3.23 This is a considerable problem since with the given sampling routine we are more likely to sample umpires who died young than umpires who died old. For this reason the t-test based on the uncensored lifetimes is not a good idea here. (It is also inappropriate to insert artificial death times for the censored group; more sophisticated techniques of *survival analysis* would be needed.)

3.23 Refer to Display 3.10.

a Yes. One should expect the rates to follow a time series where serial correlation is present.

b Here is a picture that puts them both together. There is a problem: there is a steady increase,

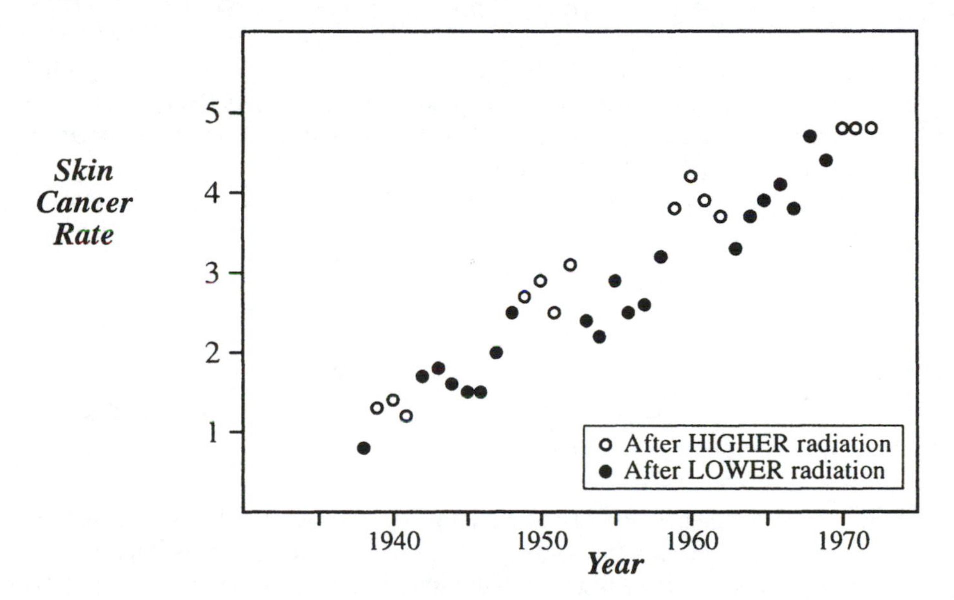

or 'trend', in the series. There is also a somewhat cyclic behavior. The trend and (possibly) the cyclic behavior are most likely unrelated to solar radiation, but they will have a strong influence on the comparison because more of the 'after higher' values fall in the later years.

3.25 Use the computer. Refer to Display 3.6.

3.27 **a** (i) Oil exporters are positively skewed. Industrialized are reasonable symmetric.

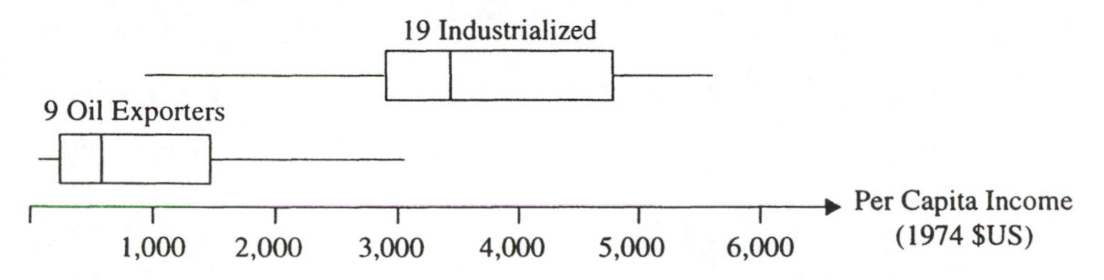

(ii) The log makes the oil exporters look OK, but the industrialized group gets squashed and has its outlier made more prominent.

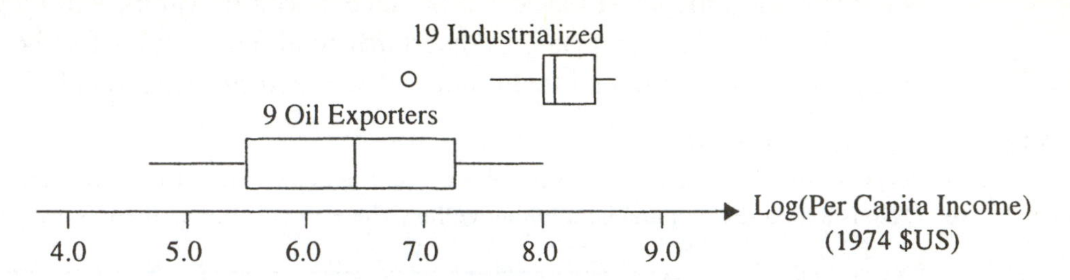

(iii) estimate = 1.703, SE = 0.280. (iv) The median per-capita income in the industrialized countries is 5.5 times that in the oil exporting companies. (v) From 1.128 to 2.278. (vi) The industrialized-to-oil exporting ratio of median per-capita incomes is estimated to be between 3.1 and 9.75 (95% confidence interval).

b (i) Oil exporters group is a bit long-tailed, and it has greater spread than industrialized group.

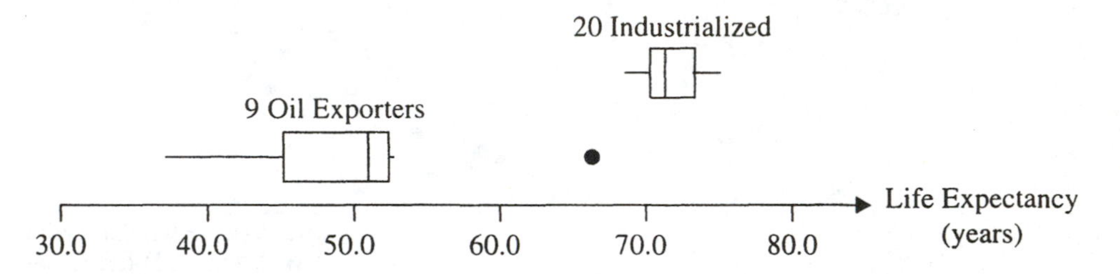

(ii) Uniform health standards; technology that insulates health conditions from environmental factors. (iii) The log, square root, and reciprocal will fail because the group with the higher average has the smaller spread. The square might work. (iv) These are the conditions where the *t*-tools fail - the group with the higher spread has the smaller sample size.

3.29 With all the data, the one-sided *p*-value is 0.0405; without the .659 value, the one-sided *p*-value is 0.0900. This is a fair swing; the evidence goes from suggestive to none.

Chapter 4: Alternatives to the *t*-Tools

4.15 One-sided *p*-value = 2/10 = .20.

4.17 **O-Ring study**. (136 + 170 + 10 + 85 + 10 + 10)/10,626 = 421/10,626 = .0396.

4.19
a 0.1718
b Normal approximation
c Continuity correction
d *t*-test gives $p = .081$; *t*-test with removal gives $p = .180$; rank sum gave $p = .1718$.
e The rank sum test is valid AND it uses all the data.

4.21 **Trauma and metabolic expenditure**. $Z = 2.95369$; two-sided p-value = .0314

4.23 **Motivation and creativity**. Two-sided p-value = .00643, compared to .00537.

4.25 **Guinea pig lifetimes**. CI: (39.59, 165.81), based on Welch's t with 93 d.f. The halfwidth is 63.11 and the critical t-multiplier is 1.9847. $SE_W = 31.80$ makes $t_W = 3.23$, giving a two-sided $p = .0016$. No. It looks like something else is involved.

4.27 **Schizophrenia Study**. Two-sided p-value = .00452, from signed rank test on the log(ratio) values. On the straight difference scale, the signed rank gives .00208 … close. It is not particularly apparent.

Chapter 5: Comparisons Among Several Samples

5.14 **Spock Trial**.

a 6.914, with 39 d.f.

b t-statistic = 5.056, with 39 d.f., giving one-sided p-value < .0001.

5.16 Use the computer.

5.18 **Fatty Acid**.

a

CPFA50	CPFA150	CPFA300	CPFA450	CPFA600	Control
168.3	171.7	146.7	151.0	152.3	185.6

Residuals vs Estimated Means:

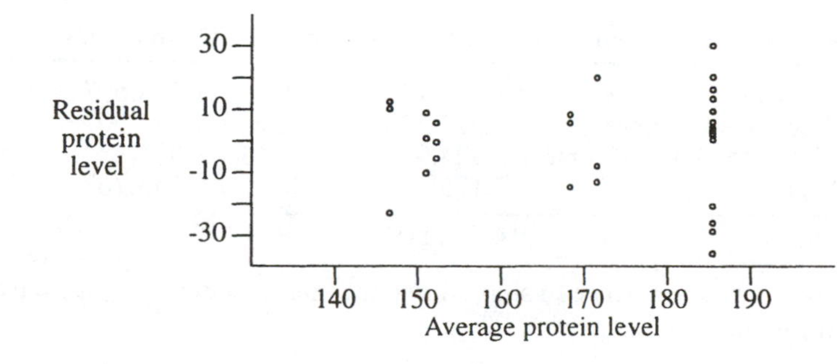

There is no suggestion that the size of residuals depends on the average protein level.
Residuals vs. Day:

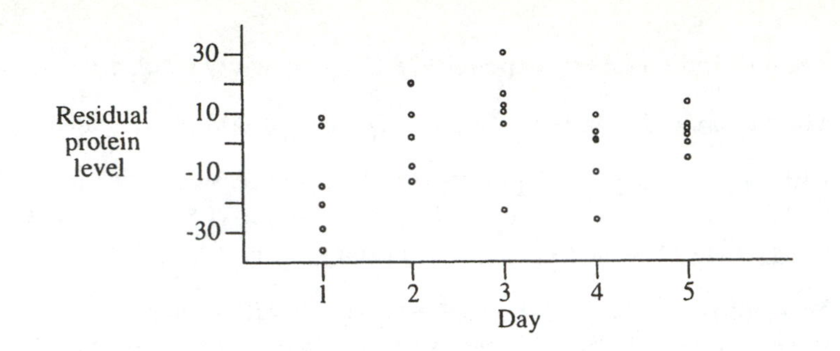

There is a suggestion that the mean level may change from one day to another.

b

	CPFA50	CPFA150	CPFA300	CPFA450	CPFA600	Control
Day 1	168.3					157.3
Day 2		171.7				195.7
Day 3			146.7			203.3
Day 4				151.0		179.0
Day 5					152.3	192.7

Analysis of Variance Table:

Source of Variation	Sum of Squares	d.f.	Mean Square	*F*-Statistic	*p*-value
Between Groups	11,147.47	9	1,238.607	7.801	.0001
Within Groups	3,175.333	20	158.767		
Total	14,322.800	29			

There is convincing evidence that the means are different.

c

Source of Variation	Sum of Squares	d.f.	Mean Square	*F*-Statistic	*p*-value
Between Groups	11,147.470	9	1,238.607	7.801	.0001
Between Treatments	7,222.533	5	1,444.507	9.098	.0001
Between Days in Control	3,924.933	4	981.233	6.180	.0021
Within Groups	3,175.333	20	158.767		
Total	14,322.800	29			

There is ample evidence to suggest that the means under the control treatment are different on different days.

5.19 Cavity Size and Use.

a 0.1919

b

Source of Variation	Sum of Squares	d.f.	Mean Square	*F*-Statistic	*p*-value
Between Groups	17.4402	8	2.1800	11.3583	<.0001
Within Groups	54.7007	285	0.1919		
Total	72.1408	293			

c The first term is $127(7.347)^2 + \ldots + 6(8.297)^2 = 16,442.9742$. The grand mean is $\overline{Y} = (127\times7.347 + \ldots + 6\times8.297)/294 = 7.4746$, so the second term (the "correction factor") is $294(7.4746)^2 = 16,425.5202$. The difference is 17.4540, which is roundoff error away from the ANOVA answer.

d You will need a between group sum of squares for the intermediate model. Following the formula in part c, you first calculate averages in the two sets of species. $\overline{Y}_1 = (127\times7.347 + \ldots + 16\times7.568)/(127 + \ldots + 16) = 1,999.4820/270 = 7.4055$; and $\overline{Y}_2 = (11\times8.214 + 7\times8.272 + 6\times8.297)/(11+7+6) = 198.0400/24 = 8.2517$. The between group sum of squares is then $270(7.4055)^2 + 24(8.2517)^2 - 294(7.4746)^2 = 16,441.3018 - 16,425.5202 = 15.7816$, with 1 d.f. (The correction factor — 16,425.5202 — is the same as in part c.) The extra sum of squares for adding different species means into this intermediate model is 17.4402 - 15.7816 = 1.6586. Putting these together as in Section 5.3.3 yields the analysis of variance table:

Source of Variation	Sum of Squares	d.f.	Mean Square	*F*-Statistic	*p*-value
Between Groups	17.4402	8	2.1800	11.3583	<.0001
Between Sets	15.7816	1	15.7816	82.2248	<.0001
Within Sets	1.6586	7	0.2369	1.2345	.2910
Within Groups	54.7007	285	0.1919		
Total	72.1408	293			

The within-sets *p*-value indicates there is no evidence that the species means are different within the two sets. There are only two different mean values.

5.21 Levene's test for the Spock trial data.

Source of Variation	Sum of Squares	d.f.	Mean Square	*F*-Statistic	*p*-value
Between Groups	41,662.54	6	6,943.76	2.0812	.0777
Within Groups	130,118.66	39	3,336.38		
Total	171,781.20	45			

The evidence suggests the possibility that the spread may differ between groups.

Chapter 6: Linear Combinations and Multiple Comparisons of Means

6.13 **Handicap Study**. For Bonferroni with only 3 groups, $k = 3*2/2 = 3$. So $[1-.05/6] = .9917$. Then $t_{65}(.9917) = 2.458$, and the SE(diff) = 0.617, giving an interval halfwidth of 1.517. The 95% confidence intervals are: (Amputee - Crutches) between -3.009 and 0.025; (Amputee - Wheelchair) between -2.431 and 0.603; (Crutches - Wheelchair) between -0.939 and 2.095.

6.15 **Comparison of Five Teaching Methods.**

a 4.484

b 1/3, -1/2, -1/2, 1/3, 1/3

c $g = 3.000$, SE(g) = 1.3645; $t_{40}(.975) = 2.021$, HW = 2.758, giving 95% confidence interval: (0.24, 5.76).

6.18 **Nest Cavities.** $g = 0.803$, $s_p = 0.4381$, d.f. = 285, SE(g) = 0.0984, HW = 0.1929. 95% CI: 0.610 to 0.996. The estimated ratio of medians is 2.23, with 95% CI from 1.84 to 2.71.

6.20 **a** $g = 0.62$; se(g) = 0.2069 => t-statistic = 2.9965; 2-sided p-value ~ 0.005

b

Differences Between Averages

	H	ML	L
ML	0.08		
L	0.10	0.02	
MH	0.13	0.05	0.03

Tukey-Kramer HSDs

	H	ML	L
ML	0.090		
L	0.108	0.112	
MH	0.077	0.083	0.102

The only difference that appears large enough to be judged real by the HSD is the H vs. MH difference.

Chapter 7: Simple Linear Regression: A Model for the Mean

7.12 **a** 15.53

b 3.94

c 4

7.14 **Planet Distances and Order from Sun.**

a Intercept estimate = 0.7648; slope estimate = 0.5369.

b & **c**

Order (X)	log(Distance) (Y)	fitted value	residual
1	1.3533	1.3017	0.0516
2	1.9782	1.8386	0.1396
3	2.3026	2.3756	-0.0730
4	2.7239	2.9125	-0.1886
5	3.3673	3.4494	-0.0821
6	3.9518	3.9864	-0.0345
7	4.5587	4.5233	0.0354
8	5.2575	5.0602	0.1973
9	5.7068	5.5972	0.1096
10	5.9789	6.1341	-0.1552

d 0.1357, with 8 d.f.

7.16 **a** **Life Expectancy and Per Capita Income.**

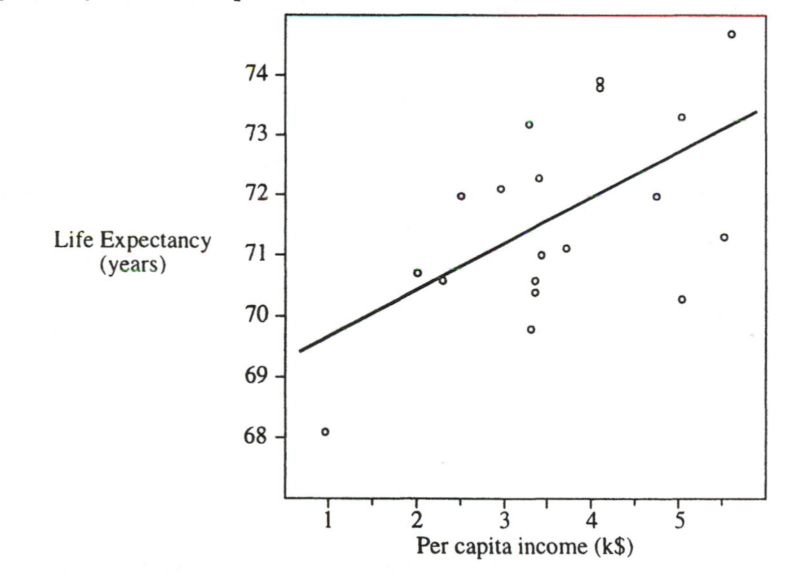

b life = 68.87 + 0.00077 income. One-sided p-value = 0.0053

c 0.572.

7.18 To calculate the right hand sides of the expressions, refer to this table:

Steer	Time	Y = pH	X = log(Time)	X^2	XY
1	1	7.02	0.0000	0.0000	0.0000
2	1	6.93	0.0000	0.0000	0.0000
3	2	6.42	0.6931	0.4805	4.4500
4	2	6.51	0.6931	0.4805	4.5124
5	4	6.07	1.3863	1.9218	8.4148
6	4	5.99	1.3863	1.9218	8.3039
7	6	5.59	1.7918	3.2104	10.0159
8	6	5.80	1.7918	3.2104	10.3922
9	8	5.51	2.0794	4.3241	11.4577
10	8	5.36	2.0794	4.3241	11.1458
	SUMS:	61.20	11.9013	19.8735	68.6928

According to the formulas, the first expression is = 19.8735 - (1/10)$(11.9013)^2$ = 5.7094; and the second is = 68.6928 - (1/10)(11.9013)(61.20) = -4.1431. Notice that these calculations can be accomplished using only five storage registers, and only one pass through the data is required. The registers keep running totals of n, Y, X, X^2, and XY. As each steer's data is entered, the totals are updated. [Pocket calculators with statistical functions use a key marked M+ or Σ+ to enter data into the registers. They also have M- or Σ- to subtract out mistakes before continuing!] Only the totals enter into the final calculations.

The left hand side expressions require a second pass through the data, because one full pass must be made to get the averages of X and Y. Therefore, either the data must be entered twice or it must be stored the first time through.

7.20 **Meat Processing**.

a SE{pred} = 0.0875

b 5.6139 < mean pH < 6.0175.

7.22 **Meat Processing**. About 109.

7.24 **a** **Crab Claw Size and Force**.

	H. nudus	*L. bellus*	*C. productus*
intercept	0.5191	-3.7800	-1.9673
SE(intercept)	1.1147	1.2842	0.9978
slope	0.4083	2.9737	2.0685
SE(slope)	0.5426	0.6125	0.4275
s_p	0.4825	0.4811	0.2981
d.f.	12	10	10

b *C. productus* vs. *L. bellus*: t-stat = 1.212. Two-sided p-value = 0.24
C. productus vs. *H. nudus*: t-stat = 2.403. Two-sided p-value = 0.025.

7.26 **a** Confirmed.

b t-statistics: -2.07, -5.71, -4.02, -5.78. The p-values are .04, < .0001, .007, and < .0001. There seems to be real evidence of a decline.

c The standard error of the estimated coefficient is quite small, so the evidence of a non-zero slope is stronger.

d It is apparent that the standard deviation about the regression, σ, is smaller.

e The proportions are based on different numbers of births. Presumably, the proportion for the U.S. is based on a larger sample size, so there is more precision in estimating the mean proportion of male births in any given year.

Chapter 8: A Closer Look at Assumptions for Simple Linear Regression

8.15 a **Island Size and Species**. Scatter plot:

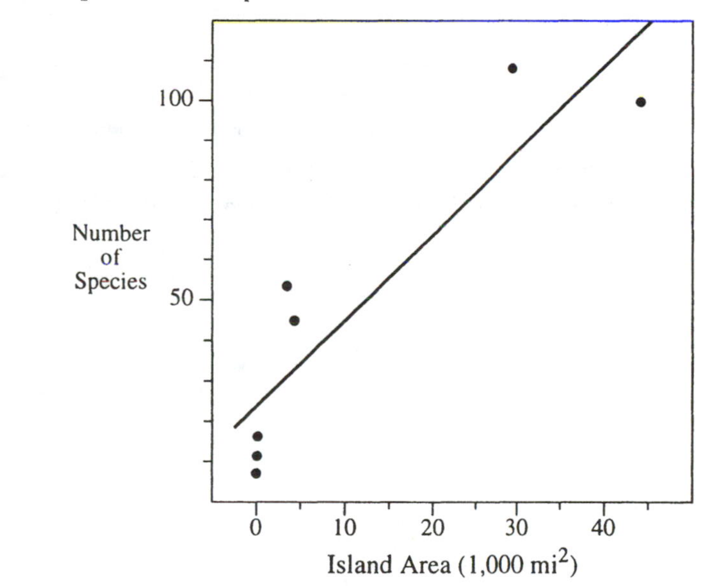

b Estimated mean number of species = 24.04928 + 0.00211×Area. The residual plot:

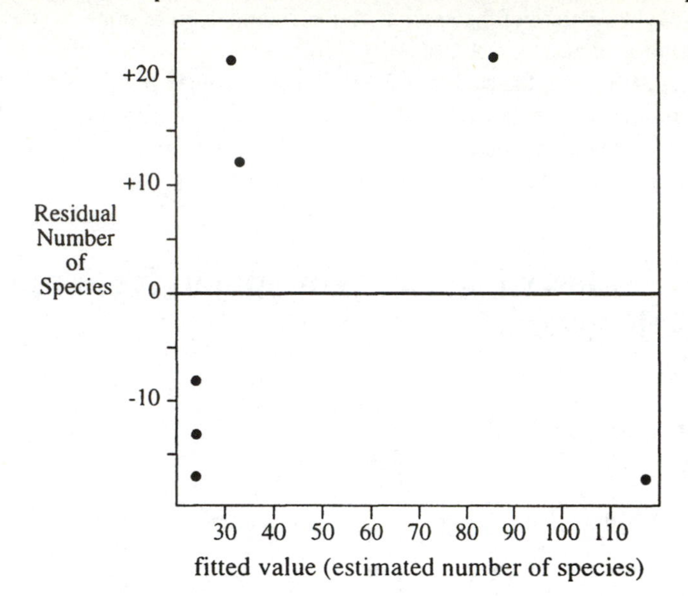

c The regression line does not come near hitting the center of the distribution of species numbers from islands with similar area. There is a pronounced curvature in the residual plot.

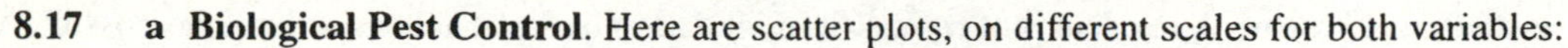

8.17 **a** **Biological Pest Control**. Here are scatter plots, on different scales for both variables:

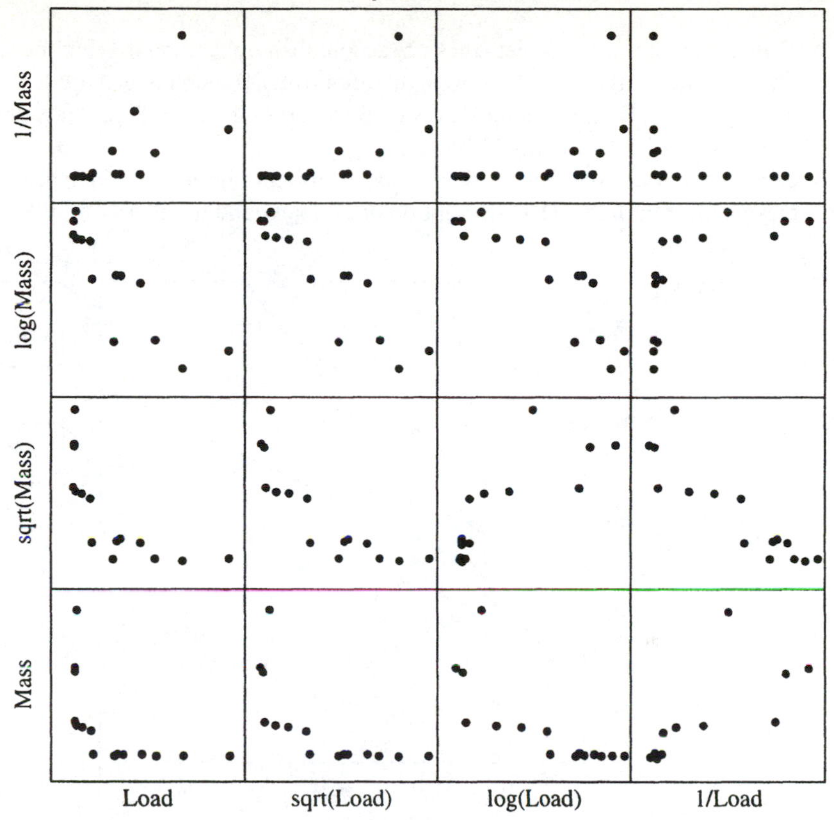

The panel most easily described by a straight line regression has log(Mass) versus sqrt(Load).

b Estimated mean log(Mass) = 3.797 - 0.262 × sqrt(Load)

c The residual plot looks satisfactory.

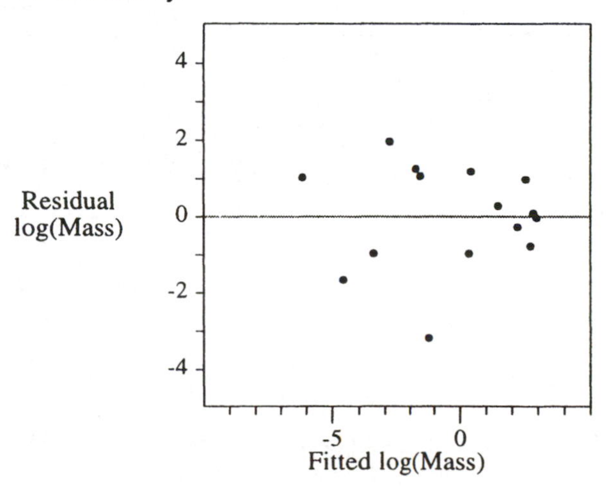

8.19 a **Pollen Removal**. A residual plot from a fit of all data shows possible curvature and possible outliers.

b Transformation of the scales does not accomplish much in clarifying the situation.

c Fits with and without the bees with duration over 30 seconds give quite different results for the regression. Examination of a residual plot from the fit without durations over 30 seconds shows no further problems.

Conclusion: For visits under 30 seconds, a straight line regression appears to give a reasonable summary. That description does not extend to visits over 30 seconds.

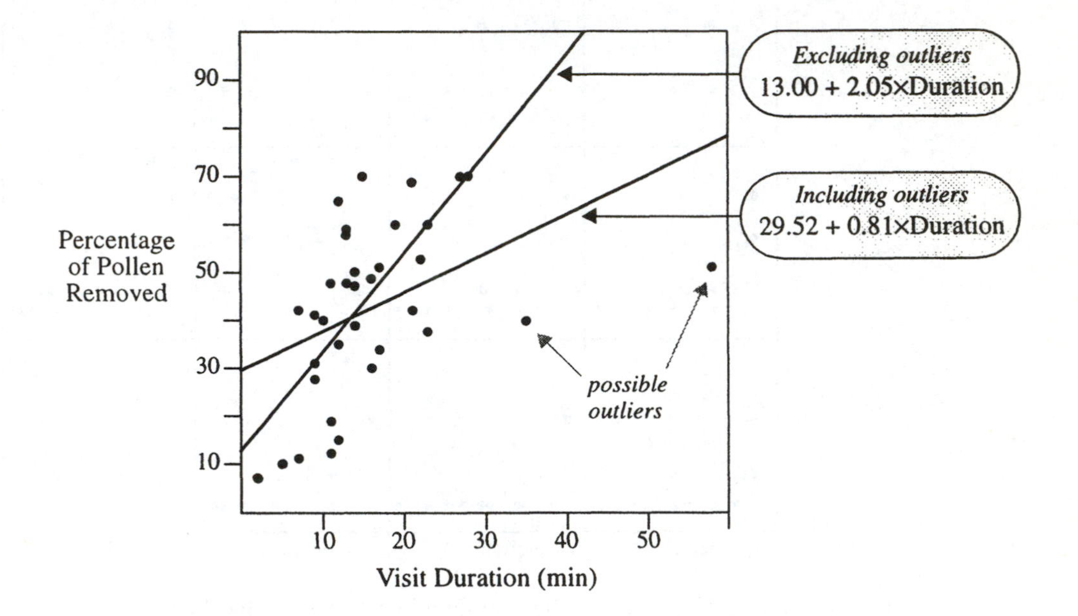

8.21 No transformation appears necessary. Case 33 and, possibly, case 35 are potential outliers.

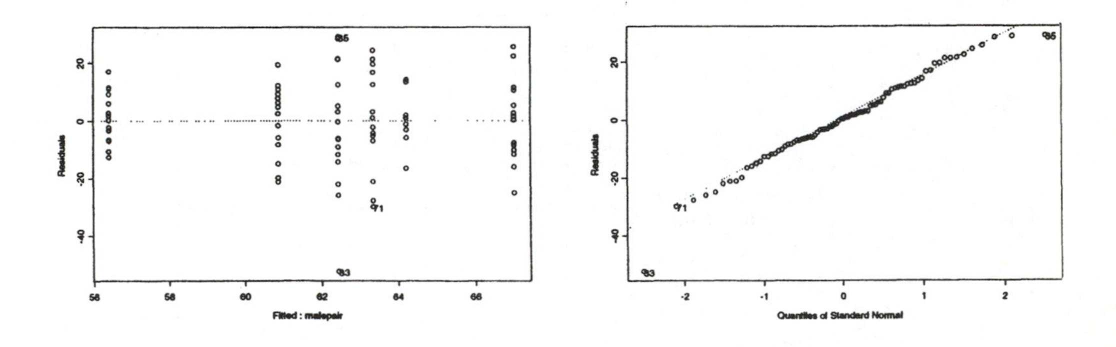

With 33 deleted: g = -21.81; se(g) = 50.62; t-statistic = -0.43

ANOVA p-value for group differences = .3744.

The conclusions remain the same with and without case 33.

Chapter 9: Multiple Regression

9.13 Meat Processing.

a The coefficient of hour-squared is 0.02243, with a standard error of 0.00681. The t-statistic is 3.29 with 7 degrees of freedom. The two-sided p-value is .0132, so there is strong evidence that the straightline relationship is not adequate for these data.

b The coefficient of the square of $log(hour)$ is -0.04463, with a standard error of 0.06245. The t-statistic of 0.71 with 7 degrees of freedom gives a two-sided p-value of .50. The straight line assumption is greatly aided by expressing hours on a log scale.

c Yes, the scale on which the squared term is insignificant might be more appropriate.

9.15 Rainfall and Corn Yield.

a 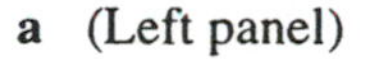
(Left panel)

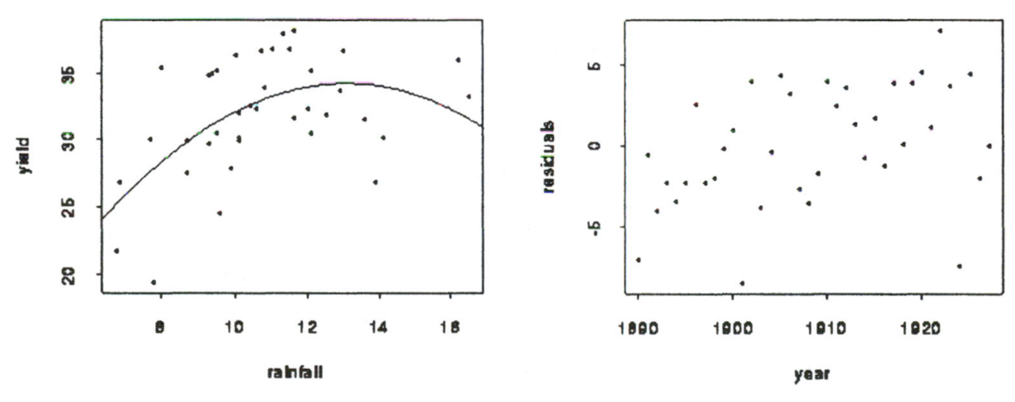

b Estimate of $\mu\{yield \mid rain, year\}$ =	-5.0 +	$6.0 rain$ -	$0.229 rain^2$	
SEs:	(11.4)	(2.0)	(0.089)	

c (See top of next page, right panel.) There is a trend; residuals tend to increase with increasing years. This means that the actual yield is larger than what is predicted from the regression of yield on rainfall for years closer to 1927 and smaller than predicted for years closer to 1890. For a given amount of rainfall, the yield per acre tended to be larger in the later years than in the earlier years. This might be due to improvements in technology.

d Estimate of $\mu\{yield \mid rain, year\}$ =	-263 +	$5.7 rain$ -	$0.216 rain^2$ +	$0.136 year$
SEs:	(98)	(1.900)	(0.082)	(0.052)
p-values:	.01	.005	.01	.01

Estimate of $\sigma\{yield \mid rain, year\}$ = 3.477 (34 d.f.)

The estimates are similar but the standard errors are a bit smaller. The estimates do not change much when *year* is added because rainfall is roughly uncorrelated with year. The standard errors are smaller, though, because σ is smaller in the model that includes the important source of variation, *year*, as an explanatory variable. As an example of the effect of rainfall, a one inch increase in rainfall from 8 inches to 9 inches is associated with an estimated 2.0 bu/acre increase in yield; but a one inch increase from 12 inches to 13 inches is associated with only a 0.28 bu/acre increase in yield.

e Estimate of $\mu\{yield \mid rain, year\}$ =

	-1909 +	159*rain* -	0.186*rain*^{2} +	1.00*year* -	0.081*rain*×*year*
SEs:	(486)	(45)	(0.072)	0.26	0.023
p-values:	.0004	.0011	.01	.0004	.0016

Estimate of $\sigma\{yield \mid rain, year\}$ = 3.028 (33 d.f.)

The two-sided *p*-value for significance of the interaction term is .0016. This indicates that the effect of rainfall on yield is smaller for years closer to 1927 than for years closer to 1890. One possible explanation is that in later years the yield became less dependent on rainfall.

9.17 Old Faithful.

a

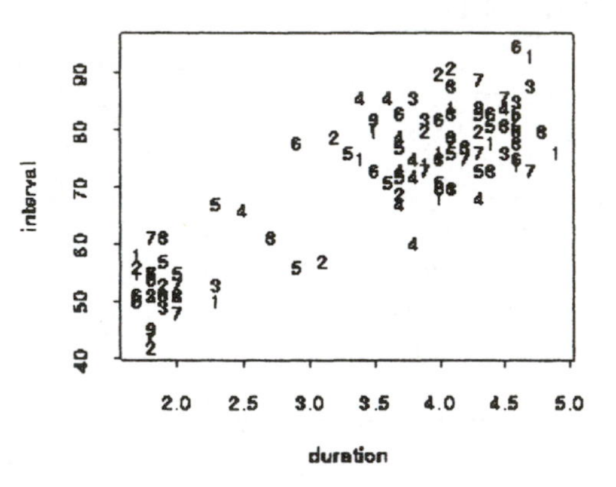

Let *datej* represent the indicator variable for date j, for j = 2, . . . , 8).

Estimate of $\mu\{interval \mid dur, DATE\}$ =

32.9 + 10.9*dur* + 1.3*date2* +0.8*date3* +0.2*date4* +0.2*date5*+2.0*date6*-0.2*date7* -0.7*date8*

(3.1) (0.66) (2.7) (2.7) (2.6) (2.6) (2.6) (2.7) (2.7)

9.19

a $\beta_0 + \beta_1 hiplus + \beta_2 hionly + \beta_3 age + \beta_4 hiplus*age + \beta_5 hionly*age$; β_5 measures divergence.

b $\beta_0+\beta_1 hiplus+\beta_2 hionly+\beta_3 age+\beta_4 hionly*age$; β_4 is the parameter of interest.

c $\beta_0+\beta_1 age+\beta_2 age^2+\beta_3 hiplus+\beta_4 hiplus*age+\beta_5 hiplus*age^2$ $+\beta_6 hionly+\beta_7 hionly*age+\beta_8 hionly*age^2$; no single parameter describes the divergence gap.

Chapter 10: Inferential Tools for Multiple Regression

10.9 **Crab Claws.**

a d.f. = *n* - #parameters = 38 - 6 = 32.

b .0014

c $t_{32,.975} = 2.037$, and the interval goes from 0.053 to 3.267.

10.11 **Butterfly Occurrences.**

a The two-sided *p*-value = .2443; the one-sided *p*-value is .2443/2 = .1222; yes. Failure to account for sampling effort could lead to concluding there was a relationship between species numbers and reserve size, when the full data would suggest that different effort may be a better explanation.

b *t*-statistic = -8.1264 with 13 d.f. The *p*-value is < .0001.

c $t_{13,.975} = 2.160$; the interval is from -0.1634 to +0.3252.

d 100 - 11.41 = 88.59%

10.13 **Bat Echolocation.**

b The slope estimate is 0.8150 for all three groups. The intercept estimates are: (i) -1.5764 for non-echolocating bats; (ii) -1.4741 for birds; and (iii) -1.4977 for echolocating bats.

c The fit summary:

Variable	Coefficient	Standard Error	*t*-Statistic	two-sided *p*-Value
CONSTANT	-1.4977	0.1499	-9.9934	<.0001
lmass	0.8150	0.0445	18.2966	<.0001
bird	0.0236	0.1576	0.1497	.8828
nbat	-0.0787	0.2027	-0.3881	.7030

d Same as b.

e Two-sided *p*-value = .8828

10.15 **Old Faithful.** The analysis of variance table for the regression of interval on duration and the day indicators:

Source of Variation	Sum of Squares	d.f.	Mean Square	*F*-Statistic	*p*-value
Regression	13,201.84	8	1,650.23	35.00	<0.0001
Residual (full)	4,620.16	98	47.14		
Total	17,822.00	106			

Dropping out the day indicators:

Source of Variation	Sum of Squares	d.f.	Mean Square	*F*-Statistic	*p*-value
Regression	13,132.99	1	13,132.99	294.08	<0.0001
Residual (reduced)	4,689.01	105	44.66		
Total	17,822.00	106			

The F-statistic for inclusion of the day indicators is $F = 0.209$ with 7 and 98 d.f. The p-value is .98 (larger than .05).

10.17 Galileo's Data

a $R^2 = 92.64\%$; adj$R^2 = 91.16\%$
b $R^2 = 99.03\%$;adj$R^2 = 98.55\%$
c $R^2 = 99.94\%$;adj$R^2 = 99.87\%$
d $R^2 = 99.98\%$; adj$R^2 = 99.95\%$
e $R^2 = 99.996\%$; adj$R^2 = 99.977\%$
f $R^2 = 100\%$; adjR^2 = undefined

10.19 Meadowfoam

a Anova table for regression of flowers on time indicator and light intensity:

Source of Variation	Sum of Squares	d.f.
Regression	3,466.700	2
Residual	871.236	21
Total	4,337.936	23

b Anova table for regression of flowers on time indicator and light intensity treated as a factor:

Source of Variation	Sum of Squares	d.f.
Regression	3682.011	11
Residual	655.925	12
Total	4,337.936	23

c $F = [(871.236\text{-}655.925)/(21\text{-}12)]/(655.925/12) = 0.437674$
From a calculator with F-percentiles, the p-value is $\Pr[F_{9,12} > 0.437674] = .89$.
Thus, there is no evidence of any inadequacy in the reduced model and therefore no evidence of lack-of-fit.

10.21

$$X = \begin{bmatrix} 1 & X_{11} & \cdots & X_{p1} \\ 1 & X_{12} & \cdots & X_{p2} \\ \cdots & \cdots & \cdots & \cdots \\ 1 & X_{1n} & \cdots & X_{pn} \end{bmatrix} \qquad Y = \begin{bmatrix} Y_1 \\ Y_2 \\ \cdots \\ Y_n \end{bmatrix} \qquad \hat{B} = \begin{bmatrix} \hat{\beta}_0 \\ \hat{\beta}_1 \\ \cdots \\ \hat{\beta}_p \end{bmatrix}$$

and

$$X'X = \begin{bmatrix} n & \sum X_{1i} & \cdots & \sum X_{pi} \\ \sum X_{1i} & \sum X_{1i}^2 & \cdots & \sum X_{1i}X_{pi} \\ \cdots & \cdots & \cdots & \cdots \\ \sum X_{pi} & \sum X_{1i}X_{pi} & \cdots & \sum X_{pi}^2 \end{bmatrix} \qquad X'Y = \begin{bmatrix} \sum Y_i \\ \sum X_{1i}Y_i \\ \cdots \\ \sum X_{pi}Y_i \end{bmatrix}$$

Then the (matrix form of the) normal equations are:

$$(X'X)\hat{B} = X'Y$$

10.23 **a** F-statistic = 8.166 on 2 and 17 d.f.; 0.01 > p-value > 0.001.
The slope difference cannot be eliminated (2-sided p-value = 0.0219); neither can the intercept difference (2-sided p-value = 0.0055). The evidence supports the existence of both differences.

b i) Yes: 2-sided p-value = 0.0032 from the estimate 2.095 with s.e. = 0.659.
ii) No: 2-sided p-value = 0.2914 from the estimate 0.707 with s.e. = 0.659.
iii) No: 2-sided p-value = 0.7302 from the estimate 0.311 with s.e. = 0.894.
iv) No: 2-sided p-value = 0.2618 from the estimate -1.754 with s.e. = 0.894.
[NB: These answers require rewriting the model using different reference levels.]

10.25 **a** Define site indicators: *s2*, *s3*, and *s4*, an irrigation indicator, *irr* and set N = nitrogen level.

b The full model is:
$\beta_0 + \beta_1 N + \beta_2 N^2 + \beta_3 irr + \beta_4 irr{*}N + \beta_5 irr{*}N^2 + \beta_6 s2 + \beta_7 s2{*}N + \beta_8 s2{*}N^2 + \beta_9 s2{*}irr + \beta_{10} s2{*}irr{*}N + \beta_{11} s2{*}irr{*}N^2 + \beta_{12} s3 + \beta_{13} s3{*}N + \beta_{14} s3{*}N^2 + \beta_{15} s3{*}irr + \beta_{16} s3{*}irr{*}N + \beta_{17} s3{*}irr{*}N^2 + \beta_{18} s4 + \beta_{19} s4{*}N + \beta_{20} s4{*}N^2 + \beta_{21} s4{*}irr + \beta_{22} s4{*}irr{*}N + \beta_{23} s4{*}irr{*}N^2$

c (i) Compare the full model with the reduction having $\beta_5 = \beta_{11} = \beta_{17} = \beta_{23}$ (= β_5, say) with an extra-sum-of-squares F-test.
(ii) Test $\beta_5 = 0$ in the reduced model from (i).
(iii) There is an intervening model with $\beta_2 = \beta_8 = \beta_{14} = \beta_{20}$ (= β_2, say). Then test $\beta_2 = 0$.

Chapter 11: Model Checking and Refinement

11.11 **Chernobyl Fallout**. Simple linear regression results are:

Variable	Coefficient	Standard Error	*t*-Statistic	two-sided *p*-Value
CONSTANT	16.726	12.420	1.347	.1980
soil cesium	0.096	0.030	3.205	.0059

Source of Variation	Sum of Squares	d.f.	Mean Square	*F*-Statistic	*p*-value
Regression	13,729.40	1	13,729.40	10.269	.0059
Residual	20,054.72	15	1,336.98		
Total	33,784.12	16			

***R*-squared** = 40.6% **adj. *R*-squared** = 36.7% **Residual SD** = 36.565

a The case influence statistics:

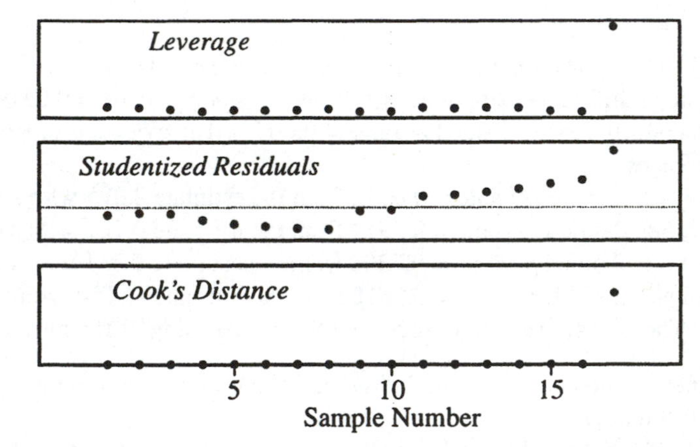

Cook's distance (the bottom line) identifies sample #17 as being highly influential. Its distance, 10.683, is huge in comparison with the distances for all other samples. This is the result of a high leverage (0.755) in combination with a large Studentized residual (3.468). Notice here that Cook's distance measures the difference in estimated parameters for the two lines fit with and without sample #17. See these lines in the solution to problem 8.18.

b After taking the logarithms of both response and explanatory variable, the case influence statistics become:

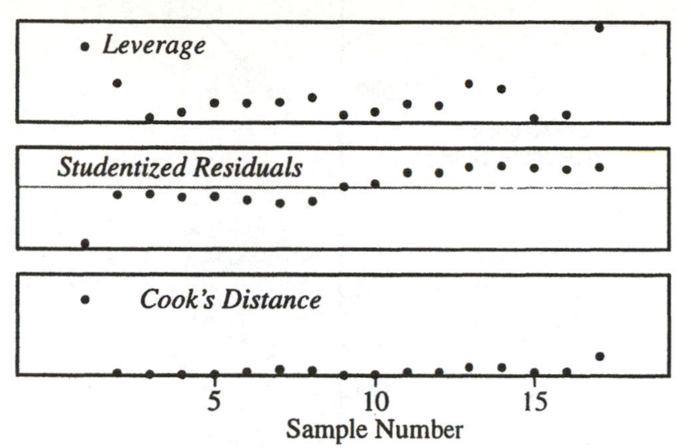

Sample #1 now becomes a problem. [As a general rule, one does not attempt (as here) to correct for a single outlier by transforming the entire data set.]

11.13 **Brain Weights**. We do, naturally. (We have large enough brains to devise a sneaky transformation to make us look good!) Although quite typical of many other mammals in its body size, gestation period, and litter size (low leverage), the human being has a Studentized residual of 3.56, reflecting its massive brain in relation to the rest of its features. The Dolphin is similarly situated, with Studentized residual of 2.52. The Tapir (-2.08) and the Hippopotamus (-2.06) have Studentized residuals that are borderline, but a few borderline residuals should be expected in a set of size 96.

11.15 **Election Fraud**.

a The picture, with the contested election #22 included, is shown on the next page.

b The fit of a regression line to the full data set gives

$$\hat{\mu}\{\text{absentee} \mid \text{machine}\} = \underset{(131.70)}{35.93} + \underset{(0.0356)}{0.0970}\ \text{machine} \qquad ; \text{and} \qquad \hat{\sigma} = 411.9$$

A listing of the internally Studentized residuals, in the same order as the data, is

Election:	1	2	3	4	5	6	7	8	9	10	11
Studres:	-1.985	-1.077	-0.636	0.306	-0.353	-1.256	0.435	0.393	0.962	0.417	-0.295

Election:	12	13	14	15	16	17	18	19	20	21	22
Studres:	0.184	-0.002	-0.167	-0.508	2.268	-0.205	-0.319	-0.432	0.124	-0.711	2.714

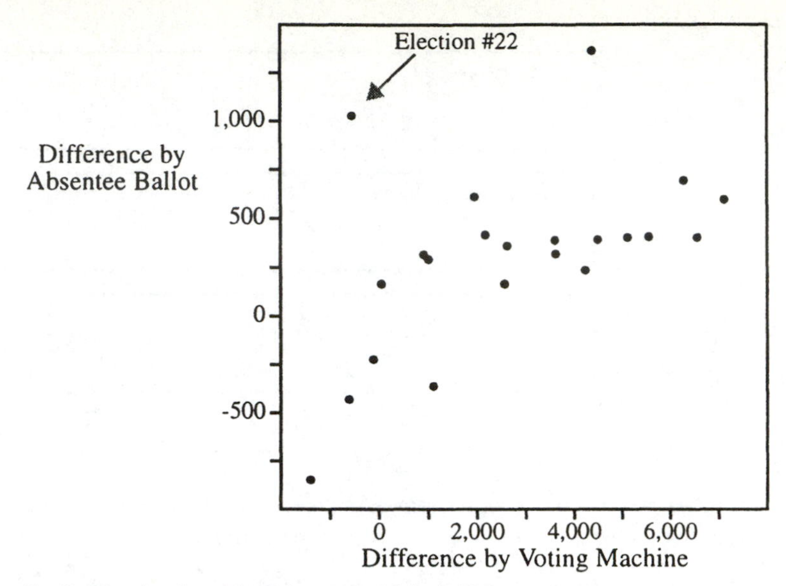

c The externally Studentized residual for case 22 is 3.327.

d The externally Studentized residual is considerably larger than the internally Studentized residual. They differ because including case 22 increases the estimate of the residual standard deviation, which is a divisor of the raw residual.

e The long-tailed aspect of these data is apparent in a normal plot of the internally Studentized residuals:

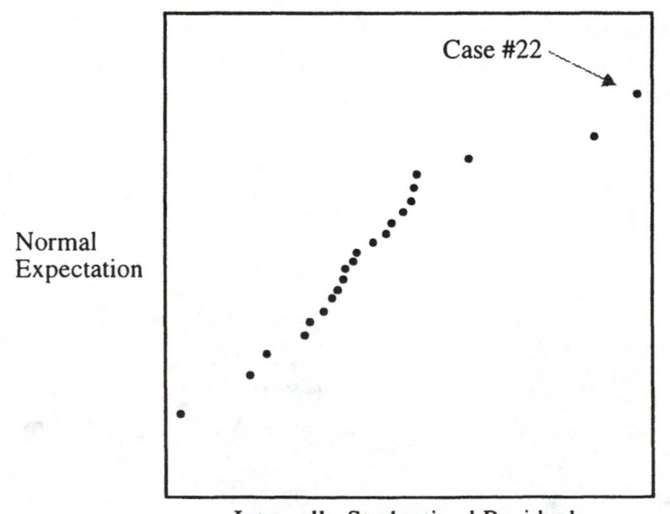

f The externally Studentized residual for case 22 (= 3.327) can be calculated as follows. First, exclude case 22, and fit a regression with the remaining 21 cases. Use this to predict the *absentee* value for case 22's *machine* value of -564. That prediction is -171.8, with a standard error of prediction = 359.7. The observed absentee value (1,025) is therefore off its prediction by [1,025 - (-171.8)]/359.7 = 3.327 standard errors. This is very large for a normal deviate and even large for a t-ratio with 19 df. However, recall that predictions are reliant on the normal distribution assumption, which is in question from other cases besides 22 in the normal plot. Because the normal plot shows that case 22 belongs to the same long-tailed pattern exhibited by the remainder of the data, there is not convincing evidence that this election was fraudulent.

11.17 **Blood-Brain Barrier**. This problem involves generating random numbers. Yours will differ from those used here, so the details may look a little different. However, the general pattern of yours should resemble the pattern in the following matrix of scatter plots.

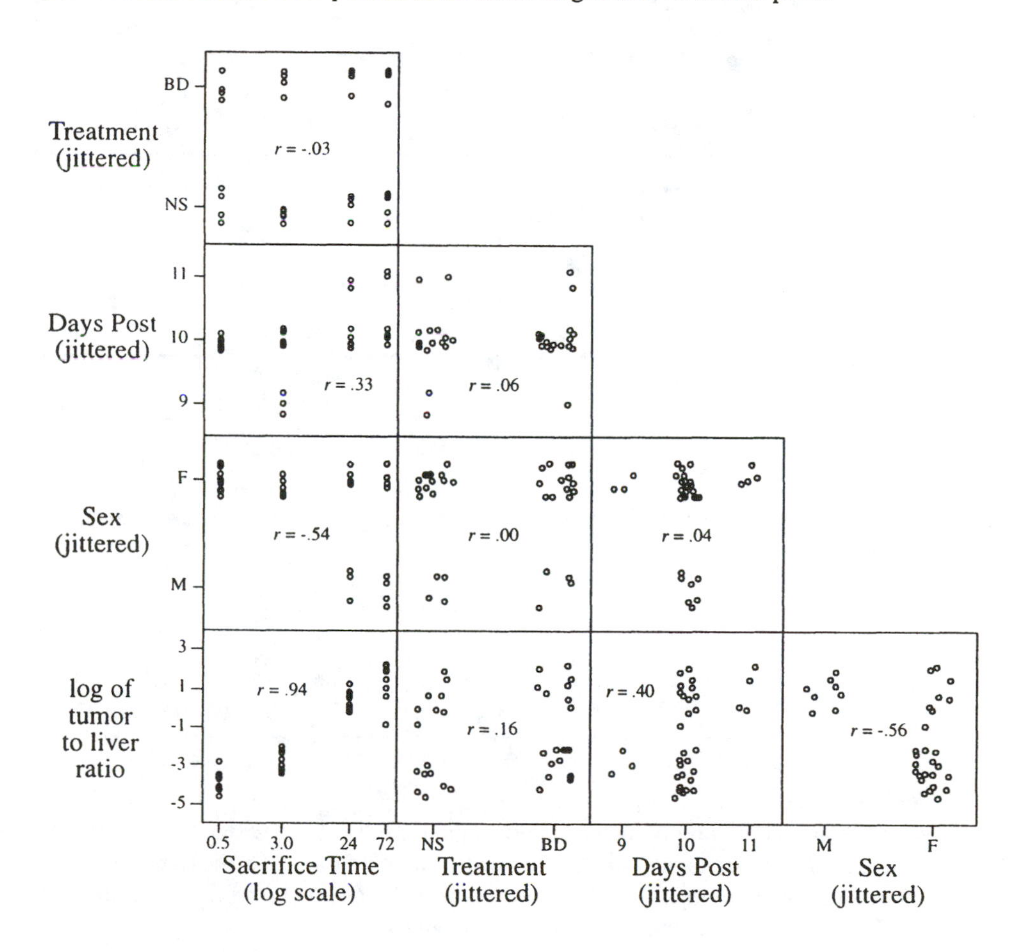

d The following relationships are apparent: sex with sacrifice time, sex with days post inoculation, and days post inoculation with sacrifice time. The corresponding correlations have the potential to cause problems with the analysis and interpretation. The strongest relationship is between the response and Sacrifice Time (lower left).

e

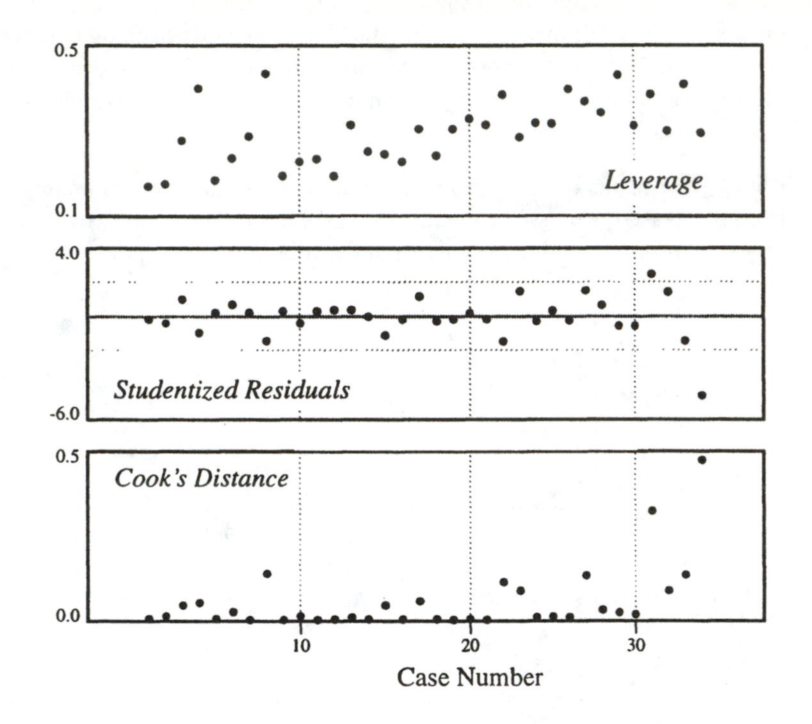

11.19 a Blood-Brain Barrier. Estimates for means of log tumor-to-liver ratio are (SEs in parentheses):

	Sacrifice Time (hours)			
	0.5	3.0	24	72
BD	-3.505 (0.195)	-2.371 (0.205)	+0.752 (0.209)	+1.649 (0.209)
NS	-4.302 (0.205)	-3.168 (0.195)	-0.044 (0.209)	+0.852 (0.209)

b Same results!

c Both are parallel regression models, in the sense that the barrier disruption minus normal saline difference is the same at all sacrifice times. Both allow the (common) pattern of means versus sacrifice time to be arbitrary (since both use three parameters to do it). These two conditions say that the two models are identical.

A more technical description goes this way. Let *BD*, *ST3*, *ST24*, and *ST72* be the indicators, so the first model looks like this:

$$\beta_0 + \beta_1 BD + \beta_2 ST3 + \beta_3 ST24 + \beta_4 ST72 .$$

Next, let a = log(0.5) = -0.693, b = log(3) = 1.099, c = log(24) = 3.178, and d = log(72) = 4.277 in the model involving X = log(sacrifice time):

$$\gamma_0 + \gamma_1 BD + \gamma_2 X + \gamma_3 X^2 + \gamma_4 X^3 .$$

Equating the means in the NS row plus the first entry in the BD row give five equations:

$$1)\gamma_0 + \gamma_2 a + \gamma_3 a^2 + \gamma_4 a^3 = \beta_0$$
$$2)\gamma_0 + \gamma_2 b + \gamma_3 b^2 + \gamma_4 b^3 = \beta_0 + \beta_2$$
$$3)\gamma_0 + \gamma_2 c + \gamma_3 c^2 + \gamma_4 c^3 = \beta_0 + \beta_3$$
$$4)\gamma_0 + \gamma_2 d + \gamma_3 d^2 + \gamma_4 d^3 = \beta_0 + \beta_4$$
$$5)\gamma_0 + \gamma_1 BD + \gamma_2 a + \gamma_3 a^2 + \gamma_4 a^3 = \beta_0 + \beta_1 .$$

Given any set of five β's in the first model, one can solve these five equations to find a set of γ's that give exactly the same values of the means to those 5 cells. Conversely, given any set of five γ's in the second model, one can solve the same equations to find a set of β's that give the same five means. But in both models, the remaining three cell means are determined by these five: the difference between the BD and NS at sacrifice time 0.5 is the same as at other sacrifice times.

11.21 **Calculus problem**. Taking the partial derivative of SS_w with respect to β_0 gives an equation which, when the summation is carried through, equals twice the difference between the left hand side and the right hand side of the first normal equation. The partial derivative of SS_w with respect to β_1 gives an equation which, when the summation is carried through, equals twice the difference between the left hand side and the right hand side of the second normal equation. And so on. So the normal equations state that all partial derivatives are set to zero, which is the condition for a critical point in SS_w.

There are several ways to argue that the solution to the normal equations is a minimum for SS_w. The easiest is simply to notice that SS_w consists of a sum of squares, each of which will get larger and larger as the proposed solution makes the fitted means further from the observed responses. So if there is a unique solution, it must be a minimum. A more difficult way is to determine the matrix of second partial derivatives for SS_w. It is twice the matrix of coefficients on the left hand sides of the normal equations. This matrix is positive definite (unless the explanatory variables are linearly dependent on each other), which is the condition for a minimum.

Chapter 12: Model Selection with Large Numbers of Explanatory Variables

12.11 At step #1, the variable with the smallest residual sum of squares is *B*. Its extra-sum-of-squares has $F = 9.217$, with 1 and 26 *d.f.* This exceeds $F_{1,26}(.95) = 4.225$, so *B* enters. At step # 2, examine models *AB* and *BC*. *AB* has the smaller residual SS, so calculate the extra-sum-of-squares *F*-statistic = 2.182 for entering *A*. This is less than $F_{1,25}(.95) = 4.242$, so *A* does not enter. Forward selection settles on the model *B*.

12.13 Here are the results of one simulation. Each solution will differ, however, in detail.

a 11.9%.

b One variable entered, using a 4.0 cutoff for the *F*-statistic.

c The Cp statistic chose a model with two variables.

d The BIC (correctly) chose a model with the constant term only.

12.15 **Blood-Brain Barrier**. Now include the treatment variable, *BD*, and three indicator variables for the three later sacrifice times.

a With the treatment variables in the model, the covariates no longer appear very important. The model with all four design variables has the smallest Cp statistic, at 3.50. The next best Cp statistic is for the same model with weight loss included. But weight loss has an F-to-enter of 2.351, and it makes the Cp statistic rise to 3.59, so it would not be included.

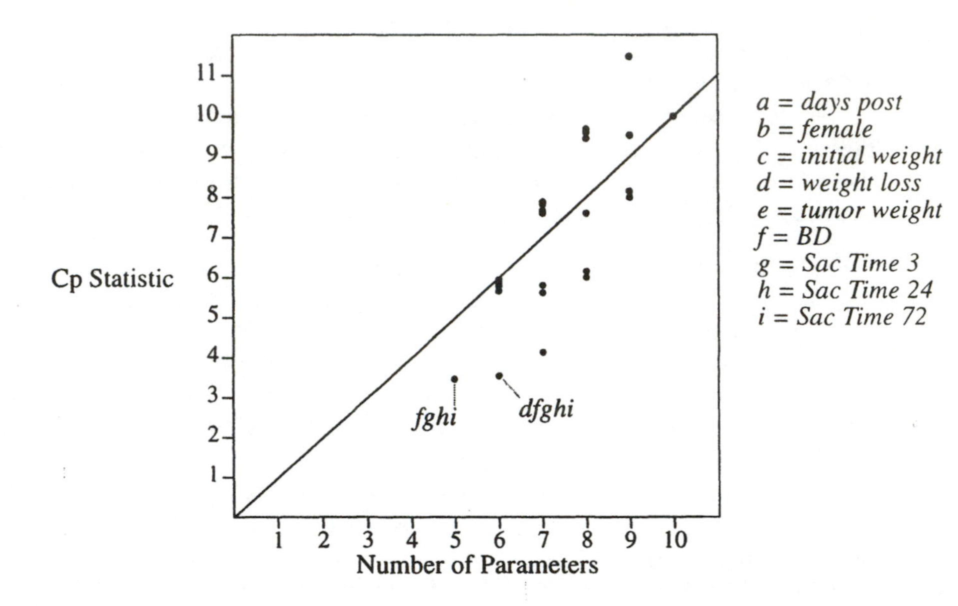

b Forward selection proceeds directly to the model with the four design variables, adding Sac Time 72 (F = 24.83), then Sac Time 24 (F = 126.28), then Sac Time 3 (F = 10.86), and then BD (F = 18.89).

c Backward elimination gets to the same model, eliminating all covariates in the order: *days post* inoculation (F = 0.01), *female* (F = 0.02), *initial weight* (F = 0.09), *tumor weight* (F = 1.66), and *weight loss* (F = 2.10).

d Stepwise does not alter the selection outcome.

e The smallest BIC (= -25.18) is for the model with only the four design variables, agreeing with the stepwise procedures. It is interesting to plot the BIC versus the number of parameters. It shows a huge gap between models with and models without all four design variables.

12.17 **Pollution and Mortality.**

a The smallest Cp statistic (3.443) goes to a model with 6 variables: *precip*, *january temp*, *july temp*, *education*, *density*, and *nonwhite*. The smallest BIC (453.3) goes to a model with 4 variables: *january temp*, *population*, *education*, and *nonwhite*. (The Cp statistic for this model is 5.49.) Adding the three pollution variables to the 6-variable model yields F = [(66,516.2 - 52,711.5)/3]/[52,711.5/50] = 4.365, with *d.f.* = 3 and 50. The p-value = .0083.

b Stepwise regression settles on the same 4-variable model identified as having the smallest BIC. Adding pollution variables, F = [(74,649.8 - 63,018.2)3]/[63,018.2/52] = 3.199, with 3 and 52 *d.f.* The p-value = .0308. The conclusion is about the same, although there is some disagreement about the strength of the evidence.

c Supposing the variables are a, b, and c, the 7 parameter (6 variable) models all include a, b, and c individually. The possible second order terms are:

1) a^2 b^2 c^2	2) a^2 b^2 ab	3) a^2 b^2 ac
4) a^2 b^2 bc	5) a^2 c^2 ab	6) a^2 c^2 ac
7) a^2 c^2 bc	8) a^2 ab ac	9) a^2 ab bc
10) a^2 ac bc	11) b^2 c^2 ab	12) b^2 c^2 ac
13) b^2 c^2 bc	14) b^2 ab ac	15) b^2 ab bc
16) b^2 ac bc	17) c^2 ab ac	18) c^2 ab bc
19) c^2 ac bc	20) ab ac bc	

Model #16, for example, is $\beta_0 + \beta_1 a + \beta_2 b + \beta_3 c + \beta_4 b^2 + \beta_5 ac + \beta_6 bc$.

12.17 Supposing the variables are a, b, and c, the 7 parameter (6 variable) models all include a, b, and c individually. The possible second order terms are:

1) a^2 b^2 c^2	2) a^2 b^2 ab	3) a^2 b^2 ac
4) a^2 b^2 bc	5) a^2 c^2 ab	6) a^2 c^2 ac
7) a^2 c^2 bc	8) a^2 ab ac	9) a^2 ab bc
10) a^2 ac bc	11) b^2 c^2 ab	12) b^2 c^2 ac
13) b^2 c^2 bc	14) b^2 ab ac	15) b^2 ab bc
16) b^2 ac bc	17) c^2 ab ac	18) c^2 ab bc
19) c^2 ac bc	20) ab ac bc	

Model #16, for example, is $\beta_0 + \beta_1 a + \beta_2 b + \beta_3 c + \beta_4 b^2 + \beta_5 ac + \beta_6 bc$.

Chapter 13: The Analysis of Variance for Two-way Classification

13.11 **Pygmalion**

a *p*-value for interaction = .72.

b The *p*-value for treatment effect in the additive model is = .012.

c Yes.

13.13 **Seaweed.** On the left is a residual plot from the fit of the log of the regeneration ratio regressed on block (as a factor), *lmp, sml,* and *big*. On the right is a plot of the estimated medians of the regeneration ratios versus block number, with treatment as plotting symbol.

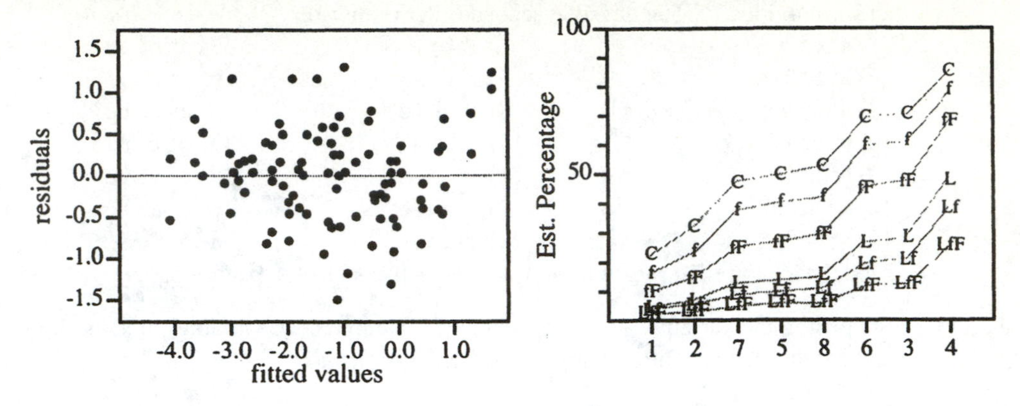

13.15 Blood-brain barrier

a Residual plot from the fit of log of blood to liver ratio on treatment and time (treated as factors) and their interaction:

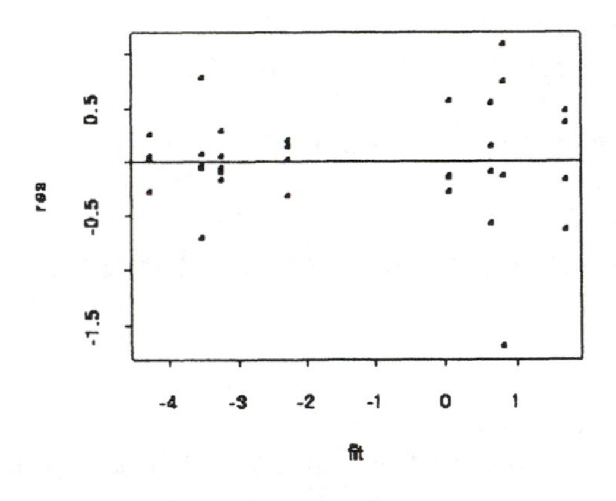

b F-statistic for interaction: 0.1672. d.f. = 3 and 26, p-value = .92. The data are consistent with there being no interaction on the log scale.

c p-value for effect of treatment = .00016. (Extra SS F-test, when sacrifice time is also in the model.)

p-value for effect of sacrifice time < .0001. (Extra SS F-test, when treatment is also in the model.)

Since the estimated coefficient of the indicator variable for the barrier disruption treatment is 0.7968, it is estimated that the median ratio of antibody concentration in the brain to that in the liver is 2.218 times as great if the disruption treatment is applied as it is if the saline solution is used.

Coefficients for the indicator variables for sacrifice times 3, 24, and 72 hours (with the sacrifice time of 0.5 hours used as a reference level) are 1.1341, 4.2573, and 5.1539 respectively. Therefore, the estimated median concentration ratio for a 3-hour sacrifice time is 3.11 times what it is at 0.5 hour. For a 24-hour sacrifice time it is 70.6 times what it is at 0.5 hour. At a 72-hour sacrifice time it is 173.1 times what it is at 0.5 hour.

Chapter 14: Multifactor Studies

14.9 Soybeans.

a Estimated mean log yield = 8.88272 + 0.76989 H_2O - 4.10881 SO_2 - 7.79585 O_3 - 4.13465 $H_2O{\times}O_3$. Estimated residual standard deviation = 0.106.

b The estimated slope against ozone is (-4.10881 - 4.13465 H_2O), and the difference between slopes at H_2O = -.05 and -.40 is -4.13465(-.05 - [-.40]) = -1.44713. The standard error of the interaction coefficient is 3.87606, so the standard error of the estimate of the slope difference is 3.87606 (.35) = 1.35662. The *t*-multiplier with 25 d.f. is 2.0595, so a 95% confidence interval for the difference between slopes is (-4.24109, +1.34683). Exponentiate to get and estimate of 0.235, with a 95% CI of (0.014, 3.845).

c Exponentiation estimates the ratio of multiplicative factors on yield associated with unit increases in ozone. Increasing ozone decreases yield (multiplicatively) under both water conditions. However, under well-watered conditions it is estimated that the multiplicative factor is only 23.5% of what it is under water-stressed conditions. (95% CI from 1.4% to 384.5%). These answers are within roundoff error of the results reported in Section 14.1.2.

14.11 **a** Solution when the die outcome is 4: treatment group = (Astrid, Ibrahim). Resulting hypothetical scores: Dawn 4.6, Astrid 3.6, Ibrahim 5.7, Michael 6.4.
Treatment group scores: Astrid = 3.6, Ibrahim = 5.7; control group scores: Dawn =4.6, Michael = 6.4. Difference in averages (control - treated) = 0.85; standard error = 1.38; *t*-statistic = 0.62.

b For coin flip = tails, heads: Astrid and Ibrahim get treated (as above, but under a different randomization scheme).

Data:

Sex	Treated	Untreated
Female	3.6	4.6
Male	5.7	6.4

The estimated coefficient of the treatment indicator variable (in the model that also includes an indicator variable for sex) is also 0.85 with a standard error of 0.15, yielding a *t*-statistic of 5.67.

c There is a much more obvious treatment effect when the sex is accounted for. Although the estimate of the effect is the same in both cases, the standard error is much smaller when sex is accounted for (in the randomized block version of the experiment in part **b**). It is smaller because SD{*score* | *treat*, *sex*} < SD{*score* | *treat*}.

14.13 Soybeans

a A paired t-test based on the straight (Forrest-Williams) difference gives an average difference of 63.63, resulting in $t = 0.50$, with 29 d.f. The two-sided p-value is .62. The Wilcoxon signed rank test for the same differences has 18 pluses and 12 minuses with the sum of the minus ranks being 177. Then $Z = -1.14$, giving a two-sided $p = .50$. On the logarithmic scale, the mean of log(Forrest/Williams) is estimated to be 0.028, giving a paired $t = 0.85$, with 29 d.f. The two-sided p-value for a test of whether the mean is 0 is 0.40. The Wilcoxon signed rank test for the logged observations has 18 pluses and 12 minuses with the sum of the minus ranks being 178. Its p-value is .52.

b The analysis begins with looking at profile plots of the response against ozone

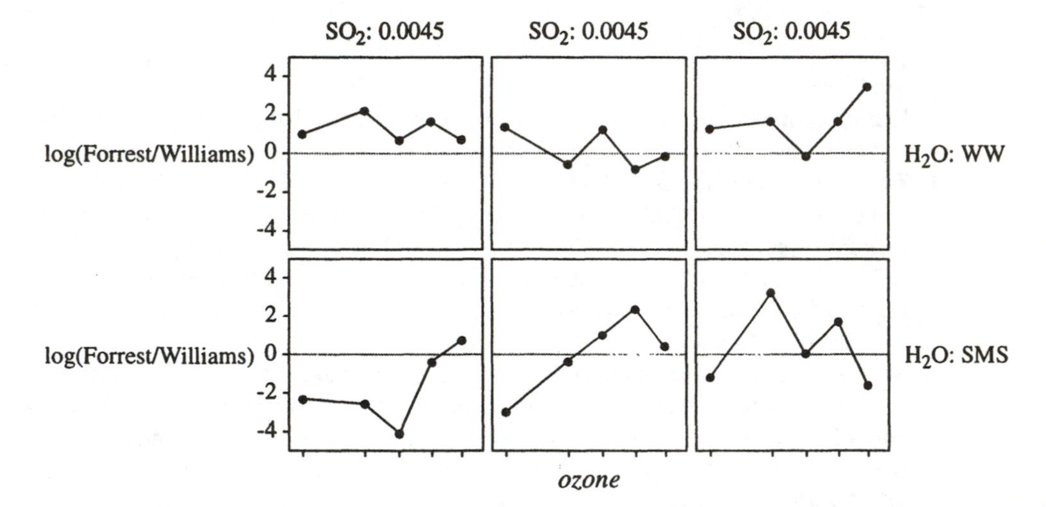

and against SO_2.

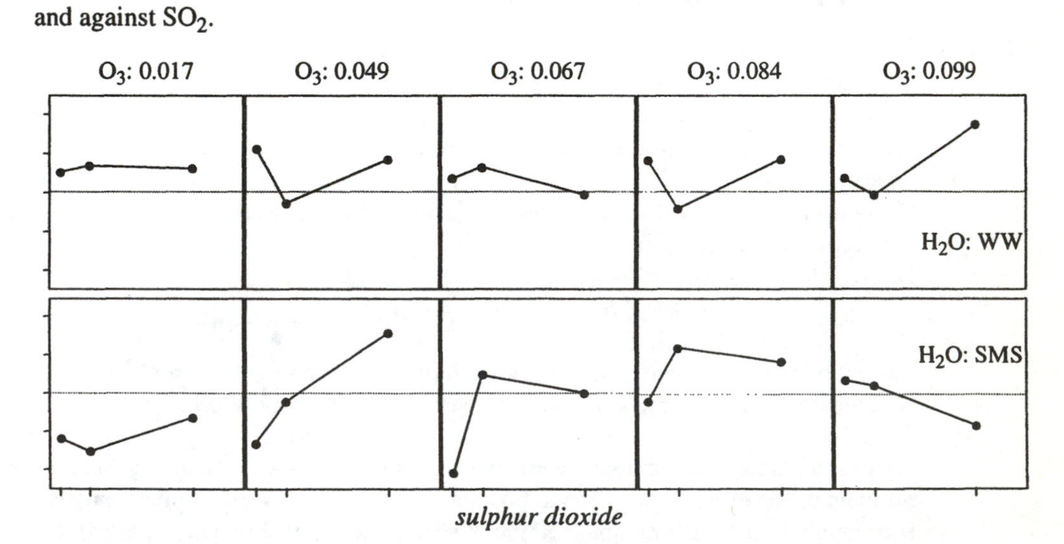

Examine these plots for consistent non-linearities (there are none apparent) and possible interactions (there is a suggestion that the slopes against ozone may be consistently different under different water conditions). The plots suggest no reason to treat SO_2 or ozone categorically or to include squared terms.

Assuming there is no three-factor interaction, the two-factor interactions were investigated, and none appeared to be significant (p-value = .44 from extra-sum-of-squares F = 0.94 with 3 and 23 d.f.). After dropping the ozone-by-SO_2 and water-by-SO_2 interactions, there is weak evidence for a water-by-ozone interaction (t = 1.47 with 25 d.f.). In the resulting additive model for inference, there is no evidence of an effect of O_3 on the relative yields (two sided p-value = .17); and there is weak evidence of an effect of SO_2 (two-sided p-value = .085). There is, however, some evidence that the relative yield of Forrest and Williams cultivars is associated with the level of water stress (two-sided p-value = .0195). The BIC chooses the model with only water stress (BIC = -100.01), whereas the Cp statistic selects a model including both water stress and SO_2 (Cp = 3.76, just 0.01 lower than a model including all three main effects and the ozone-by-water stress interaction). In the model with only water stress, it is estimated that the median ratio of Forrest to Williams yield is increased by 15.6% (i.e. changes by a factor of 1.156) when the chambers are subjected to a soil moisture stress of -0.40 MPa. A 95% confidence interval on this factor is formed on the log scale from the estimate 0.1453 and its standard error, 0.0584. A generous multiplier should be used. Because the original model contained 6 variables and had 23 d.f., a reasonable multiplier would be the square root of $6F_{6,23}(.96)$, or 3.894. The interval would be: (-9.0%, +47.0%).

Chapter 15: Adjustment for Serial Correlation

15.7 Old Faithful.

a If your computer does not compute the number of runs directly, here is one method. First, define an indicator variable (***plus***) for positive residuals. Its sum is the number of positive residuals (p = 47), and 107 minus its sum is the number of negative residuals (m = 60). Next, create a lagged version, *plus*(-1) of the indicator variable. The absolute difference, *diff* = |*plus* - *plus*(-1)| is an indicator variable for when a run ends, so the sum of *diff* is the number of runs, 66. From Section 5.4.2, the expected number of runs is μ = 53.71, and the standard deviation is σ = 5.07. The test statistic, with continuity correction, is Z = 2.325. The two-sided p-value, .0201, strongly suggests serial correlation is present. Since there are more runs than expected, the serial correlation is negative — above average residuals tend to be followed more often by below average residuals, and belows followed by aboves.

b The first serial correlation coefficient is r_1 = -0.3610. [Calculation of this coefficient may be done in a slightly different way by different computer packages. Some use the full series average to compute both residuals and lagged residuals, while others correct using only the averages of the series used in forming the estimate. The difference is slight.] The quick standard deviation is 0.0967, so Z = -3.73 gives a two-sided p-value = .0002.

c A summary of the indicated regression:

Variable	Coefficient	Standard Error	*t*-Statistic	two-sided *p*-Value
CONSTANT	58.5931	6.5260	8.98	<.0001
duration	9.1665	0.7697	11.91	<.0001
lag duration	3.2570	1.1316	2.88	.0049
lag interval	-0.4308	0.0946	-4.47	<.0001

Estimate of residual standard deviation = 0.5136

The first serial correlation coefficient for the residual series from this regression is r_1 = 0.0017, which is not significantly different from zero (two-sided *p*-value = .9856).

15.9 Sunspots.

a A good strategy for plotting time series is to experiment with several choices of scale on both the time and the response axes. Different scales emphasize different features. Usually, however, the most informative scales have long time axes relative to the range of response. This is the case here.

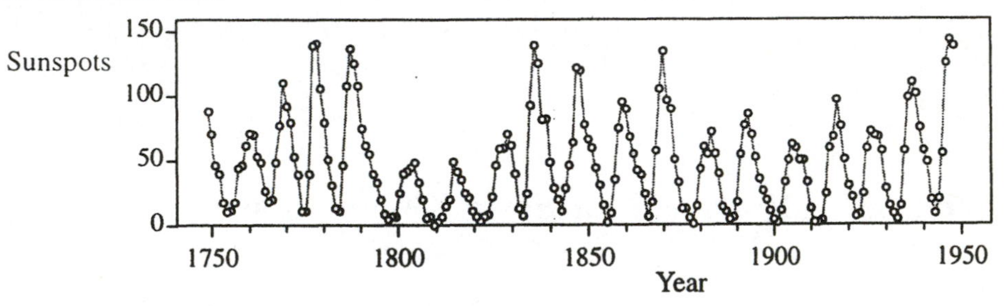

The troughs appear more rounded than the peaks, which suggests a transformation. The logarithm over-transforms, whereas the square root produces a reasonable symmetry.

b After taking the square root transformation of the sunspot numbers, a picture of the partial autocorrelation function estimate is as follows:

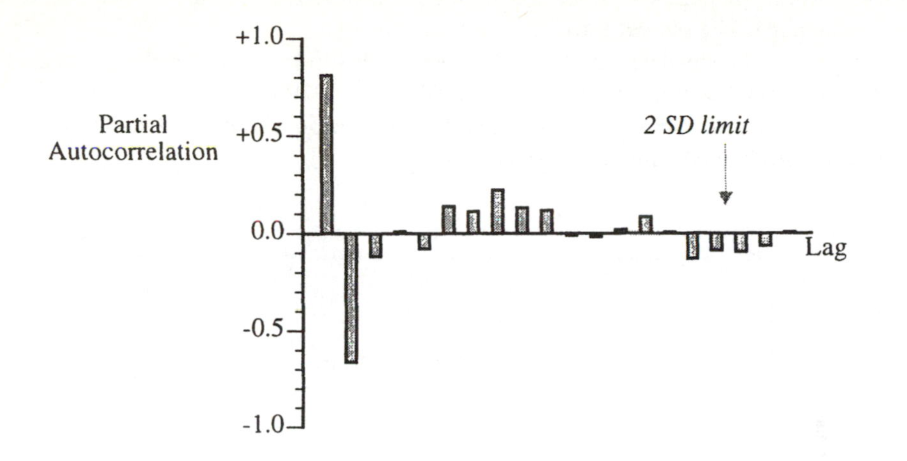

There is clear evidence that at least a second order autoregressive model is required. Some larger lag correlations are near or exceed 2 standard deviations from zero, so there is some concern that a higher order model may be necessary. [1 SD = 1/√200 = 0.0707.] The issue is resolved by appealing to the Bayes Information Criterion:

Lag	PACF	Residual Variance	BIC	Posterior Probability
0	—	1.0000	5.30	.0000
1	.8109	.3424	-203.77	.0000
2	-.6636	.1916	-314.93	.9298
3	-.1160	.1891	-311.94	.0698
4	.0092	.1890	-306.66	.0004
5	-.0786	.1879	-302.60	.0000
6	.1364	.1844	-301.06	.0000
7	.1129	.1820	-298.33	.0000
8	.2211	.1731	-303.05	.0000
9	.1302	.1702	-301.17	.0000
10	.1171	.1679	-298.64	.0000
11	-.0116	.1678	-293.36	.0000
12	-.0197	.1678	-288.14	.0000
13	.0198	.1677	-282.92	.0000
14	.0822	.1666	-278.98	.0000
15	.0055	.1666	-273.69	.0000
16	-.1295	.1638	-271.77	.0000
17	-.0885	.1625	-268.05	.0000
18	-.0976	.1610	-264.66	.0000
19	-.0637	.1603	-260.18	.0000
20	.0038	.1603	-254.88	.0000

[Note: The response variance, 7.6780, is the residual variance for lag zero. It appears in the first BIC formula on page 442, but is a constant that is dropped from the second BIC formula on the same page. The second formula was used above.]

The smallest BIC (-314.93) corresponds to the second order autoregressive model, which may be afforded a posterior probability of .9298.

Conclusion: For this segment of the sunspot series, an AR(2) model for the square root transformed numbers appears satisfactory.

15.11 **a**

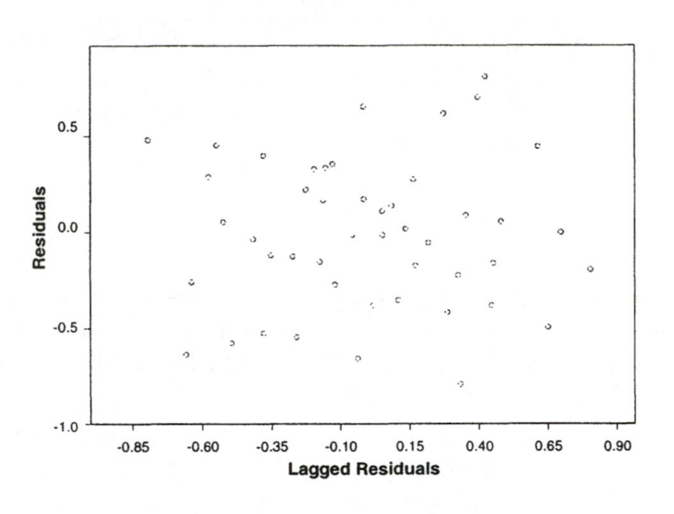

b

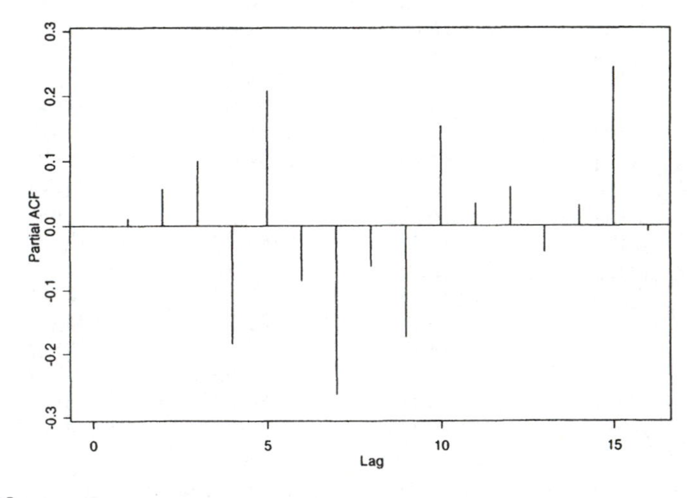

c No.

Chapter 16: Repeated Measures

16.5 **Nature-Nurture.** The normal expectation would be that the average IQ of a child would be related to the natural mother's IQ, *BMIQ*, and that the change in IQ would be more closely connected to the foster mother's education level, *AMED*. The research questions concern whether there are additional relationships involved.

a The correlation between residuals from regression, $r = 0.1310$, reflects the lack of any noticeable relationship in their scatter. This suggests that separate analyses are permitted.

b (i) R-squared = 11.08%, with adjusted R-squared = 9.60%. This is not great, but the two-sided p-value is .0082, giving strong evidence for the relationship between average IQ and *BMIQ*. (ii) $p = .7379$ shows no evidence that *AMED* contributes anything. (iii) R-square = 0.49%, with a negative adjusted R-squared. The p-value here of .5888 shows no evidence that the change in IQ is linearly related to the foster mother's educational level. (iv) $p = .0153$ gives strong evidence that there is some relationship between the change in IQ and the natural mother's IQ.

16.7 For (first primary, third primary), the pooled SDs are (8.3136, 3.9315); the standard errors for the species differences are (3.0004, 1.4189); and the pooled correlation is $r = .4947$.

a First primary: $t = 2.1397$ has two-sided p-value = .0409. Third primary: $t = 2.1918$ has two-sided
p-value = .0366. Because the 97.5th percentile in the t-distribution (29 df) is 2.0452, neither 95% confidence interval includes zero.

b T-squared = 6.2788; $F = 3.0311$ on 2 and 28 d.f.; so p-value = .0643.

c The resulting 95% confidence interval includes (0, 0), because the p-value exceeds .05.

d This may also seem to be contradictory, but only to those who cling doggedly to the .05 level cutoff. The evidence registered against zero values — .0409, .0366, .0643 — is relatively consistent.

16.9 **Memory data**. Here are four examples representing different regions around the confidence ellipse.

Short term	Long term	T-square	F-statistic	p-value
8	8	13.76	6.45	.0095
11	-7	7.80	3.65	.0509
25	0	5.61	2.63	.1049
0	0	26.30	12.33	.0007

16.11 Religious competition.

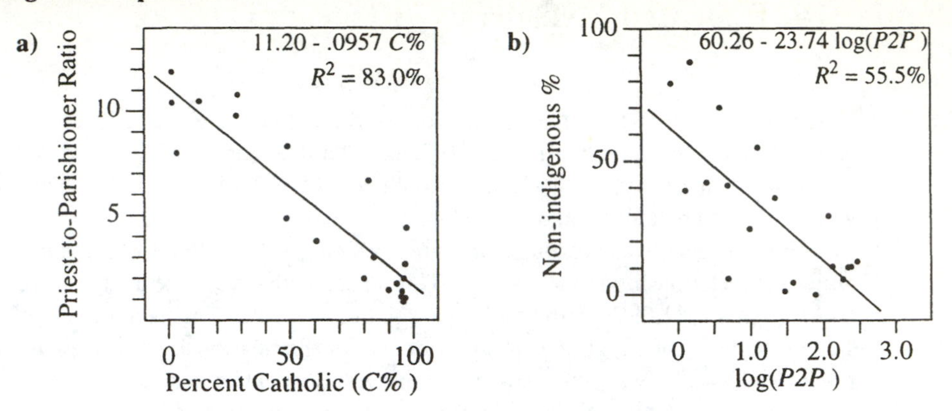

These regressions exhibit the most prominent relationships between the three variables.

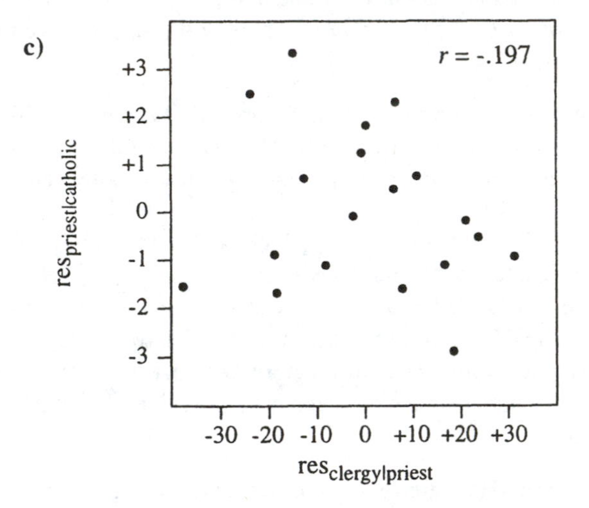

d (i) A multivariate regression is the best way to view the situation, with the *P2P ratio* and the *%Indigenous* as responses to *%Catholic* as the explanatory variable. (ii) The inference from multivariate regression can be approximated by separate inferences from univariate regression when the correlation between residuals from the univariate regressions is small. The choice of separate regression equations in this problem was made to capture the most prominent relationships between the variables. The choice succeeds in producing residuals with small correlation. (iii) Countries with high population percents being Catholic generally have low numbers of priests per parishioner, and the countries with low numbers of priests per parishioners generally have higher percents of the priests being non-indigenous. These are descriptive features (supporting the hypothesis), and the usual cautions concerning observational data apply.

Chapter 17: Exploratory Tools for Summarizing Multivariate Responses

17.9 **Pig fat**. The first three principal components account for about 95% of the total variation (which is 1,268.0438). The complete breakdown is:

Component	Variance	Percent (%)	Cumulative %
PC1	1,096.8540	86.50	86.50
PC2	55.4819	4.38	90.87
PC3	46.8717	3.70	94.57
PC4	25.4784	2.01	96.58
PC5	21.1703	1.67	98.25
PC6	13.7165	1.08	99.33
PC7	4.1800	0.33	99.66
PC8	2.1819	0.17	99.83
PC9	1.6478	0.13	99.96
PC10	0.4422	0.03	100.00
PC11	0.0191	0.00	100.00
PC12	0.0000	0.00	100.00
PC13	0.0000	0.00	100.00

The coefficients for the first three principal components are:

Variable	PC1	PC2	PC3
M1	.2606	.3488	-.0752
M2	.2070	.8034	.2223
M3	.3572	.0108	-.0679
M4	.3223	.0090	-.0981
M5	.2946	.0117	.4065
M6	.2922	.0727	-.3344
M7	.2857	-.1233	-.5546
M8	.3152	-.1406	-.0024
M9	.3251	-.0838	-.1814
M10	.2585	-.1239	.2534
M11	.2866	-.3261	.4036
M12	.1638	-.1778	-.0013
M13	.1523	-.1786	.2932

PC1 is nearly the average of all MRI readings. PC2 is dominated by M2. PC3 contrasts the average of M5 and M11 with the average of M6 and M7. Depending on how expensive MRI measurements are, the following sets of linear combinations may be considered as surrogates for:

(1) LC1 = (M2+M5+ M6+M7+ M11)/5; LC2 = M2; and LC3 = (M5+M11)/2 - (M6+M7)/2; or

(2) LC1 = (M2+M5+M7)/3; LC2 = M2; and LC3 = M5 - M7.

17.11 a **Love and marriage.** The largest canonical correlation between passionate responses and compassionate responses is .5433. The test that all four canonical correlations are zero has a *p*-value of .4375, indicating no evidence of any correlation between the two sets of responses.

b The largest canonical correlation between husbands' responses and wives' responses is .5717, with associated *p*-value = .4572. Again, there is no evidence that the two sets of responses are correlated in any way.

Chapter 18: Comparisons of Proportions or Odds

18.9 a **Heart Disease and Obesity.** (i) 0.01832 in obese group; 0.01537 in not obese group. (ii) 0.00505. (iii) -0.00696 to +0.01285.

b *z*-statistic = 0.5825; one-sided *p*-value = .2801.

c (i) 0.01866 and 0.01561 (ii) 1.195 (iii) 0.304 (iv) 0.66 to 2.17.

d The odds of CVD death in the obese group are estimated to be between 0.66 and 2.17 times the odds of CVD death in the not obese group (95% confidence interval).

18.11 a (i) 0.00050 (ii) 0.000250 (iii) 0.6429

b (i) 0.00025 (ii) 0.00025 (iii) 0.4737

c (i) 0.00025 (ii) 0.00025 (iii) 0.1692

Retrospective samples (of equal size) do not estimate the population proportions.

18.13 a If, for example, $\pi_u = .0010$ and $\pi_v = .0002$, the relative risk is $\rho = 5.0$, while the odds ratio is $\omega = 5.004$.

b 3.69

c 2.403

Chapter 19: More Tools for Tables of Counts

19.11 $C_{503,0} \times C_{317,5} / C_{820,5} = (317 \times 316 \times 315 \times 314 \times 313)/(820 \times 819 \times 818 \times 817 \times 816) = .0085$.

19.13 a Excess = 5.5; SE(Excess) = 1.963; *z*-statistic = 2.80; one-sided *p*-value = .0026.

b Fisher's exact test: one-sided *p*-value = .0044.

19.15 a No. The expected number of used and not used houses of old wood are both smaller than 5.

b One-sided *p*-value = .0057.

Chapter 20: Logistic Regression for Binary Response Variables

20.9 a **Donner party.** Estimated survival probabilities for males are .413 at age 25 and .091 at age 50. For females, the estimates are .777 at age 25 and .332 at age 50.

b Set the estimated log-odds to zero and solve for age. For females, the age of 50% survival is 41.0 years; for males it is 20.5 years.

20.11 **a** **Space shuttle**. The fitted logistic regression model for the probability, π, of failure given temperature is:

$$\text{logit}(\hat{\pi}) = \underset{(5.7031)}{10.8753} - \underset{(0.0834)}{0.1713}\ \text{temperature}$$

b One-sided p-value = .0200, from z-statistic = -2.054.
c Drop in deviance = 5.9441 with 1 d.f. gives p-value = .0148. This is a two-sided p-value for the coefficient, so the approximate one-sided p-value based on the deviance would be .0074.
d The 95% confidence interval extends from -0.3348 to -0.0078.
e logit = 5.5650; estimated probability of failure = .9962.
f It represents an extrapolation beyond the range of the available data.

20.13 **Donner party**. The deviance for the reduced model is 14.8734, with 14 d.f.; the deviance for the full model is 13.8026, with 13 d.f. The 1 d.f. drop in deviance is 1.0708, which gives a p-value of .3008. There is no evidence that 30 years is not the correct value.

Chapter 21: Logistic Regression for Binomial Counts

21.9 **Moth coloration**. The method outlined in this problem is a main-effect way of seeing how the relative numbers of the two morphs taken change with distance. Contrast this with the method of the text, where the desired effect was an interaction. [Caution: the method only works when equal numbers of the two morphs are placed out at each location.]

a At Sefton Park, for example, a total of 31 moths were removed: 17 were light (Typical), and 14 were dark (Carboneria). The logit requested is log(17/14) = 0.1942. Continuing, one arrives at this plot (next page):

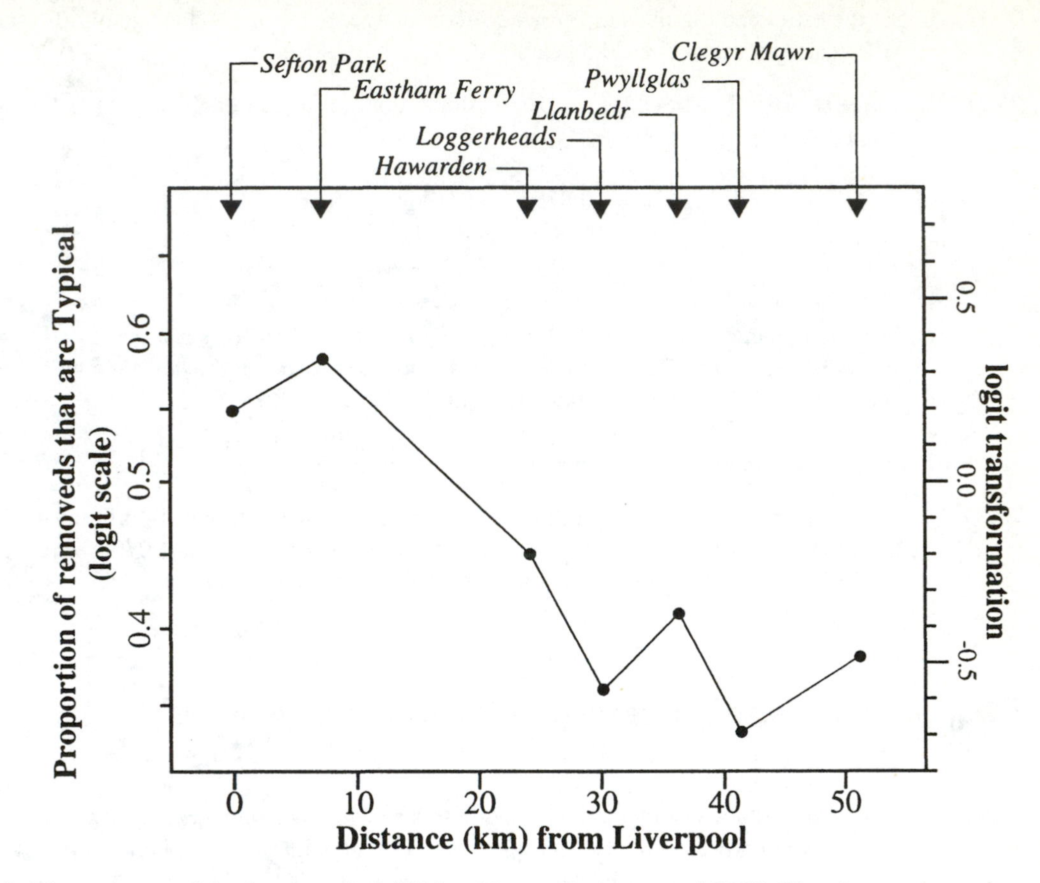

b The estimate of the intercept is 0.2805, with standard error = 0.2309. The slope estimate is -0.0187, with standard error 0.0068.

c The deviance statistic is 2.07, with 5 d.f. The p-value for goodness of fit is .8388, so there is no evidence for lack of fit.

21.11 With aggravation level as a categorical factor represented by indicators $L1, \ldots, L5$ for the first five levels and having level 6 as its reference level, the logistic regression fit is:

$$\text{logit}(\hat{\pi}) = 27.30 - 30.72\,L1 - 29.11\,L2 - 27.33\,L3 - 26.22\,L4 - 24.84\,L5 - \underset{(0.5426)}{1.7409}\,black$$

The drop in deviance for interaction terms is 2.2391, with 5 d.f. (p-value = .8152). There is no evidence to suggest that the equal odds ratio assumption is inadequate.

21.13 **Vitamin C.**

a The empirical logits in the four groups are: -1.205, -1.245, -1.178, and -1.319, respectively.

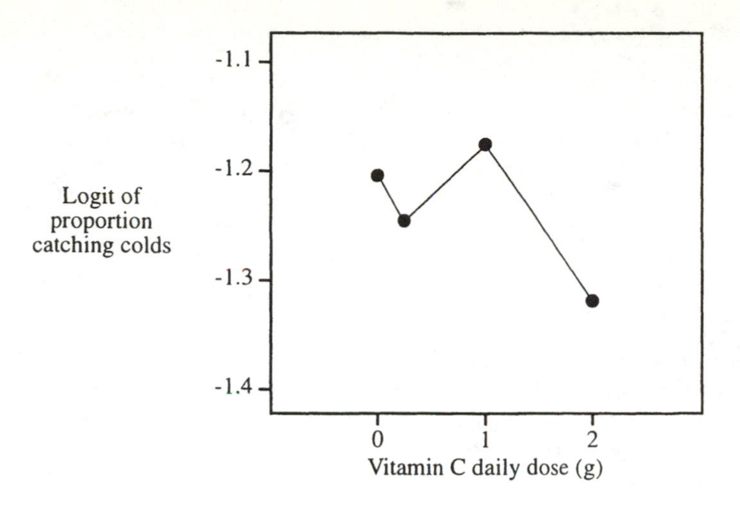

b The fit:

$$\text{logit}(\hat{\pi}) = -\ 1.2003 - 0.0346\ \text{dose}$$
$$(0.0617)\quad (0.0711)$$

The goodness of fit p-value = .7674, from the deviance = 0.5296 with 2 d.f. Wald's test for the coefficient of dose is z = -0.4866, giving a two-sided p-value = .6265. The drop in deviance test has a p-value = .6253, based on the drop = 0.2385 with 1 d.f.

c There is not evidence that the model is inadequate. Nor is there much evidence that the odds of a cold are associated with the daily dose of vitamin C.

Chapter 22: Log-Linear Regression for Poisson Counts

22.15 **Elephant Mating**

a Estimate of $\mu\{matings^{1/2} \mid age\} = -0.8122 + 0.0632 age$

b Estimate of $\mu\{\log(matings + 1) \mid age\} = -0.6989 + 0.0509 age$

c Estimate of $\mu\{matings \mid age\} = \exp(-1.582 + 0.06869 age)$

d Lines representing the three fits are shown on the scatterplot below. There are no obvious inadequacies with models **a** and **b**, although **c** is probably the easiest to interpret.

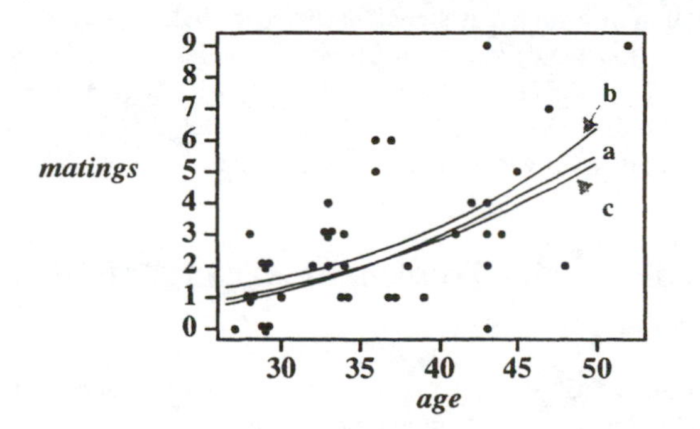

22.17 **Obesity and heart disease**. The two-sided p-value from the deviance test is .73.

22.19 **Galapagos Islands**

- **a** The deviance goodness of fit p-value is less than .0001, providing overwhelming evidence of lack of fit of the Poisson model.
- **b** Using quasi-likelihood analysis and backward elimination—discarding a variable at each step if its t-statistic is smaller in magnitude than 2—leads to the model with log area and log of area of the nearest neighbor for describing the mean number of non-native species.
- **c** For each doubling of island area the mean number of nonnative species increases by 35%. For a given island area, the mean number of nonnative species decreases by 9% for each doubling of area of the nearest island.

22.21 **El Nino and Hurricanes**.

- **a** There is convincing evidence that the mean number of storms is associated with El Nino temperature, after accounting for West African wetness (2-sided p-value = .004 from a likelihood ratio test). It is estimated that the mean number of storms is reduced by 17% when the El Nino temperature changes from cold to neutral or from neutral to warm (95% confidence interval 6% to 26%).
- **b** There is convincing evidence that the mean number of hurricanes is associated with El Nino temperature, after accounting for West African wetness (2-sided p-value = .007 from a likelihood ratio test). It is estimated that the mean number of hurricanes is reduced by 20% when the El Nino temperature changes from cold to neutral or from neutral to warm (95% confidence interval 6% to 31%).

Chapter 23: Elements of Research Design

23.13 **Taxol**. Here the response is binary: Y = 1 for responding to the drug and = 0 for non-response. The control proportion is 0.60 for the standard treatment, and the desired alternative is 0.75 for taxol. The desired odds ratio is 2.0. The sample size in each group should be 304.

23.15 **Infantile paralysis**. The non-victim proportion is 0.20, corresponding to odds of 0.25. An odds ratio of 3.0 translates to odds of 0.75, or a victim proportion of 0.4286. Distinguishing the two can be done with a sample of 132 from each group.

23.17 **Big Bang**. According to the Big Bang Theory, the age of the universe is the slope in the regression of distance on velocity. The half-width of the 95% confidence interval for slope is $t_{.975}(n-2)\times\sigma/[(n-1)S_X^2]^{1/2}$, where σ is the standard deviation about the regression and S_X is the sample standard deviation of the explanatory variable. With $\sigma = .25$ and $S_X = 400$, the question requires the calculation of n such that the *width* (twice the half-width) is 0.001021. Using $t_{.975}(n-2) \cong 2$, $2\times2\times.25/[(n-1)400^2]^{1/2} = .001021$, which implies that n = 7. Further refinement shows that with n=7, $t_{.975}(5) = 2.571$, and the CI width is .00131. With n=8, $t_{.975}(6) = 2.447$, and the CI width is .00116. So the answer is 8. When S_X is taken to be 600 the answer is n=7 (at n=6 the width is .001034, just a bit wider than what is required).

Chapter 24: Factorial Treatment Arrangements and Blocking Designs

The exercises ask the students to generate their own random numbers. The following solutions use computer generated random numbers for illustration.

24.11 **a** In the following, each group of four leaves represents one full replicate.

LEAF	TRTMT	RESPONSE	A	B	AB	L2	L3	L4	L5	L6	L7	L8	L9	L10	L11	L12
1	(A, -)	1.890	1	0	0	0	0	0	0	0	0	0	0	0	0	0
1	(A, B)	0.618	1	1	1	0	0	0	0	0	0	0	0	0	0	0
2	(- , -)	-1.027	0	0	0	1	0	0	0	0	0	0	0	0	0	0
2	(-, B)	1.319	0	1	0	1	0	0	0	0	0	0	0	0	0	0
3	(A, B)	1.439	1	1	1	0	1	0	0	0	0	0	0	0	0	0
3	(-, B)	-1.501	0	1	0	0	1	0	0	0	0	0	0	0	0	0
4	(A, -)	1.156	1	0	0	0	0	1	0	0	0	0	0	0	0	0
4	(- , -)	0.908	0	0	0	0	0	1	0	0	0	0	0	0	0	0
5	(- , -)	-0.516	0	0	0	0	0	0	1	0	0	0	0	0	0	0
5	(A, -)	0.483	1	0	0	0	0	0	1	0	0	0	0	0	0	0
6	(- , B)	-0.165	0	1	0	0	0	0	0	1	0	0	0	0	0	0
6	(A, B)	-0.615	1	1	1	0	0	0	0	1	0	0	0	0	0	0
7	(- , -)	-0.496	0	0	0	0	0	0	0	0	1	0	0	0	0	0
7	(- , B)	-2.371	0	1	0	0	0	0	0	0	1	0	0	0	0	0
8	(A, B)	-0.833	1	1	1	0	0	0	0	0	0	1	0	0	0	0
8	(A, -)	1.054	1	0	0	0	0	0	0	0	0	1	0	0	0	0
9	(A, -)	-1.039	1	0	0	0	0	0	0	0	0	0	1	0	0	0
9	(A, B)	-0.059	1	1	1	0	0	0	0	0	0	0	1	0	0	0
10	(A, -)	-0.832	1	0	0	0	0	0	0	0	0	0	0	1	0	0
10	(- , -)	0.289	0	0	0	0	0	0	0	0	0	0	0	1	0	0
11	(A, B)	1.540	1	1	1	0	0	0	0	0	0	0	0	0	1	0
11	(- , B)	0.106	0	1	0	0	0	0	0	0	0	0	0	0	1	0
12	(- , -)	-0.849	0	0	0	0	0	0	0	0	0	0	0	0	0	1
12	(- , B)	-0.414	0	1	0	0	0	0	0	0	0	0	0	0	0	1

b The responses appear above. There are indicator variables for A, for B, for the interaction of A and B, and for 11 of the 12 leaves (blocks). The resulting regression fit is: 0.715 +0.616A -0.272B +0.119AB -0.433L2 -0.841L3 +0.009L4 -1.039L5 -1.200L6 -2.013L7 -1.144L8 -1.803L9 - 1.294L10 +0.013L11 -1.211L12. Adding 6 to both responses in leaf 8, 2 to both responses in leaf 11, and 15 to both responses in leaf 3 does not change coefficients of A, B, and AB. The coefficients of L3, L8, and L11 increase by 15, 6, and 2.

24.13 **a** Use the Latin square in Display 24.10. Generate 64 responses, corresponding to the cells in the table. Represent treatments by 7 indicator variables for the numbers 2 through 8 appearing in the cells of the square. Represent columns by 7 indicators for columns 2 through 8; and represent rows by 7 more indicators for rows 2 through 8. This results in the analysis given in Display 24.11.

b Add 6 to all responses in row 1, 12 to all responses in row 2, and so on. Perform a similar operation on the responses in columns. As stated, the analysis of treatment effects is not changed by this operation.